Praise for *Wild Plant C*

Jared Rosenbaum explains how our yards, farms, and cities can be optimized for abundance, diversity, and resiliency by the planting of edible, natural systems. A brilliant approach to ecological restoration simultaneous with food production!

— Mark Shepard, author,
Restoration Agriculture: Real World Permaculture for Farmers

Jared Rosenbaum's book has arrived at the right time. Understanding how modern European agriculture is ecologically illiterate is the starting point to creating respectful, innovative ecosystems restoration worldwide. Although the plant species in this book are North American, the concepts Jared describes are globally relevant and eye-opening to anyone new to this work. Australian ecosystem restoration, known as Bushcare or Dunecare, have been happening for years. Having worked in this space, I am in full agreement with Jared's work.

— Rosemary Morrow, deep green teacher, refugee supporter,
and author, *Earth Restorer's Guide to Permaculture*

Jared tempts us to explore a multiplicity of tastes, textures and colors, locally sourced, tended by human care and at a human scale. This thoroughly researched and deeply conceived book helps relieve our disconnection from the natural world, leads to the preservation of irreplaceable plant communities, and reacquaints us with the singular delectable tastes that can only be provided by the plants of our native landscapes.

— Gerould Wilhelm, co-author,
Flora of the Chicago Region: A Floristic and Ecological Synthesis

This book will give new sight to the land and plants around you. It is a cornucopia of useful information, from soils and land use history to plant inventories and their invaluable uses, the tools to caretake them and the inspiration to continually seek to learn more.

— Kelly Kindscher, ethnobotanist, University of Kansas,
author, *Medicinal Wild Plants of the Prairie*

Rosenbaum's work is an uncommonly thorough reference book, a primer on plant ecology, restoration biology, and the medicinal and edible properties of the Mid-Atlantic plants from his region. Much more than that, it is thought-provoking, aspirational and the first chapter alone is worth the price of the book. Jared is taking a risk here. He is hoping that we are smart enough to recognize and act on the wisdom clearly outlined in these pages. I trust that he is right.

— Doug Tallamy, entomologist, ecologist, conservationist,
and author of several books, including *Bringing Nature Home*

This gorgeous and well-researched book is a delightful must-read for any person interested in restoration of native habitats in cities and beyond, the stories the land tells us, and the edible and medicinal uses embodied in the wild plants around us.

— Dr. Lena Struwe, director, Chrysler Herbarium, Rutgers University

Jared Rosenbaum explores the fascinating connections between people and their natural environments. It is a well crafted and important work for anyone who seeks to serve them both.

— Larry Weaner, Larry Weaner Landscape Associates

This guidebook holds one of many pathways to healing the ecological wounds of colonialism. Rosenbaum outlines a practice that is both revolutionary and ancient: tending native plant communities for the simultaneous benefits of human welfare and wildness. Drawing from his deep knowledge of plant ecology and restoration, Jared points us towards long-term, fulfilling relationships with the natural communities we are entrusted to care for. This message could not be more timely.

— Sam Thayer, author, *The Forager's Harvest*

WILD PLANT CULTURE

A Guide to Restoring
Edible and Medicinal Native Plant Communities

Jared Rosenbaum

To elves
May your native
garden grow, a haven
for creatures large +
small and a balm
for your soul.
With love
Jared

new society
PUBLISHERS

Cover design by Diane McIntosh.

Cover Image: © Jared Rosenbaum.

Printed in Canada. First printing October 2022.

Inquiries regarding requests to reprint all or part of *Wild Plant Culture* should be addressed to New Society Publishers at the address below.

To order directly from the publishers, order online at www.newsociety.com

Any other inquiries can be directed by mail to: New Society Publishers

P.O. Box 189, Gabriola Island, BC V0R 1X0, Canada (250) 247-9737

LIBRARY AND ARCHIVES CANADA CATALOGUING IN PUBLICATION
Title: Wild plant culture : a guide to restoring edible and medicinal
    native plant communities / Jared Rosenbaum.
Names: Rosenbaum, Jared, author.
Description: Includes bibliographical references and index.
Identifiers: Canadiana (print) 2022026662X | Canadiana (ebook)
    20220266638 | ISBN 9780865719804 (softcover) | ISBN
    9781550927733 (PDF) | ISBN 9781771423694 (EPUB)
Subjects: LCSH: Restoration ecology—Canada, Eastern. | LCSH:
    Restoration ecology—East (U.S.) | LCSH: Wild plants, Edible—
    Canada, Eastern. | LCSH: Wild plants, Edible—East (U.S.) | LCSH:
    Medicinal plants—Canada, Eastern. | LCSH: Medicinal plants—East
    (U.S.) | LCSH: Human ecology.
Classification: LCC QK98.5.N6 R67 2022 | DDC 581.6/3—dc23

Funded by the Government of Canada   Financé par le gouvernement du Canada

New Society Publishers' mission is to publish books that contribute in fundamental ways to building an ecologically sustainable and just society, and to do so with the least possible impact on the environment, in a manner that models this vision.

Contents

Acknowledgments

THANK YOU to everyone who read and commented on the manuscript, from its early inscrutable days to its nearly finished form. Karl Anderson, Kerry Barringer, Chris Berry, Pat Coleman, Kerry Hardy, Roger Latham, Leslie Sauer, Daniela Shebitz, Lena Struwe, and Mike Van Clef, your expertise and friendship is inspiring and deeply appreciated.

Special thanks to Kerry Barringer for being my colleague, friend, and mentor across thousands of acres of fieldwork together.

To Kelli Kovacevic and Matt Trump at Morris County Park Commission, thanks for enabling so much of the botanical survey work that informed this book.

To everyone at New Society Publishers who worked with me on this book, it's been amazing to be part of the team. Caylie Graham, thanks for reading the unsolicited manuscript and championing it.

Thank you to the plant people for everything you give, and what you have asked in return.

Lots of love to Rachel and Beren, always.

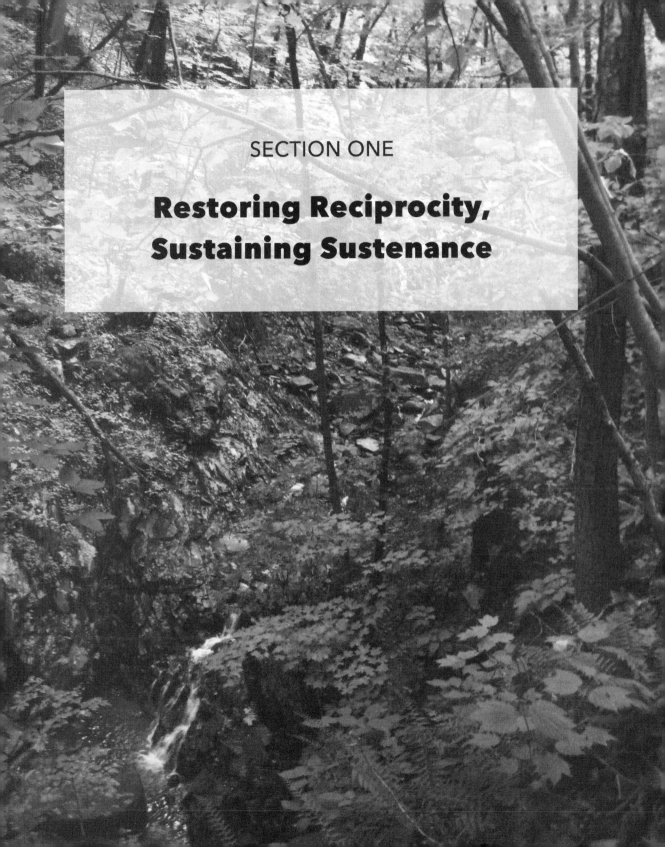

SECTION ONE

Restoring Reciprocity, Sustaining Sustenance

Introduction

ABOUT A DECADE AGO, someone wrote a book about the island where I grew up. They described its 30 species of native orchids, glacial bogs, and old growth forests of oak, hickory, and chestnut along its hilly shoreline. This remembrance of my childhood island brought a tear to my eye — because the island where I grew up is Manhattan, and the natural habitats described in the book were from over 400 years ago.

We all live in habitats. Some of us live in habitats that are buried beneath pavement, some in habitats diminished by centuries of intensive land use. Some of us have yards and farms that contain remnant habitats of exceptional quality.

We live in habitats, but are unaccustomed to thinking of them that way. We've drawn a hard line between what is natural and what is human, to the detriment of all. We need human areas — urban and otherwise — that welcome wildlife, *and natural areas that welcome humans.*

How can we break down the divide between human-occupied spaces and what is "natural"? As direct participants in natural communities — as creatures who find food, medicine, and purpose in natural areas, whether those are our backyard, a city park, a farm, or a nature preserve.

This is a book about restoring your habitat: to abundance, beauty, function, and utility as well. For all wildlife — including us two-leggeds. Restoring habitats using edible and medicinal wild plants, in particular, welcomes us back to the natural community, because these species feed and heal us just like they do the other animals.

At the heart of ecological restoration is a reciprocal exchange. In exchange for the sustenance plants give us, we offer our human skills to restore and repair degraded plant habitats. Ecological restoration is a way of connecting to and healing nature, returning the gifts that we receive from the plants and the living earth.

This book will help you match plants to habitats, and plants to people. It is built around the idea of the plant community. If you find a place with one or two plants present, these indicate that other plants, other members of that community, may also thrive there. In addition to describing plant communities, this book illustrates the personality of each plant species — how they might fit into the community, if you will.

The time has come to reconnect with our habitats, right where we live, work, and play. Not as museum pieces, but as vital, sustaining elements of our lives, livelihoods, and lifeways.

The Farmer's Quandary

Recently I was at a small gathering of organic farmers at a friend's farm in central New Jersey, about an hour away from where I live. They were talking about their farm soils and how they are working to improve them. We were gathered in a circle in a friend's open pole barn, the mid-October evening chill working its way through our light autumn garb with every gust.

Many farmers there described composting, cover cropping, and no-till farming, and then the woman in front of me began to speak. Her farm was in the sandy coastal plain of New Jersey. She described how the soil pH started at around 4 (very acidic), and through copious application of lime they had brought it up to well over 7 (near

neutral, but much more alkaline than most native soils). Her farm had been a forest when they bought it, so they cut it down, pulled all of the stumps, and began to amend and alter the soil. Now she grows organic vegetable crops there.

I'm not here to second-guess her progressive farming practices, but to raise a question. Is there another way to produce food, medicine, and other economically and culturally important plant materials — without tearing down and replacing the natural habitats found in one's region? Because all farms, no matter how regenerative, were once natural habitats.

We modern humans are constantly seeking that perfect "river valley" soil — deep and loamy, neutral pH, highly fertile — wherever we garden or farm. Yet few of us live in natural soils that have this character, so we try to create it with whatever inputs and manipulations we can.

Affordable land here in heavily populated New Jersey is often marginal from a farmer's perspective — poorly drained clays or rocky hillsides. Much of the prime farmland has been developed or fetches a premium price — that farmland which once earned New Jersey the moniker "The Garden State." Often young farmers and homesteaders end up contending with conditions that are anathema to annual vegetable production, adding tons of compost and other inputs in an effort to "fix" clay or rocky soils. Another farmer

Picking fruit in the Pine Barrens.

at the same little gathering had dumped over a hundred tons of mushroom compost on two acres of land in a quest to make it arable.

Let's return for a moment to our coastal plain farmer, the one who started with a soil pH of 4. The coastal plain features sandy, highly drained soils that in former epochs were oceanfront and undersea. These silica-based sands leach nutrients and are generative of soils low in pH, and they tend to be very well drained. Not so great from a conventional horticultural perspective. However, the natural plant community in the coastal plain is wildly abundant with fruit- and nut-bearing woody plants, tea herbs, greens, and

medicinal roots and flowering tops. What would it look like to work with the existing native plant communities that our landscapes do or could support? That is what this book is about.

Two Human Paths

To speak simply, human existence can be divided into two different economic lifeways that spring from two very different food production systems.

The first is what wild foods author Sam Thayer has artfully dubbed "ecoculture."[1] Some may be more familiar with terms like "hunter-gatherer," though this term seems inadequate to reflect the reciprocal tending of the non-human world found in these cultures. In ecoculture, we are keystone animals, participating in and constantly melding the natural ecology around us to be its most abundant and productive. Ecoculture management practices benefit an entire plant and animal community, and we humans thrive on the resulting abundance. This lifeway rewards awareness and deep knowledge of ecology, and produces cooperative relationships, including among humans, who typically live in egalitarian societies. The path of ecoculture is typical of many human cultures for the past 10,000 years, probably much more.

The second path is a type of agriculture based around monocrops of domesticated plants, usually annuals. Here, humans clear away natural communities and optimize conditions for a narrow group of crop species intended to benefit ourselves exclusively. Nature becomes vilified as the weeds, pests, predators, and weather patterns that constantly threaten our fragile domesticates. Annual agriculture has brought about cultures deeply suspicious of nature, and societies that are hierarchical and characterized by servitude and exploitation.

The above two paragraphs can hardly do justice to the spectrum of human experience. Yet they lay a framework and pose a question. Our exploitative modern economy, rooted in monocrop agriculture, is leading to a dead end for nature and natural humans. Can we look to our other human lifeways such as ecoculture for answers on how to better structure our food production and our relationships with the natural world and each other?

This book is about practicing ecoculture in northeastern North America. Rooted in wild plant communities and the practice of ecological restoration, it explores food and medicinal plant species as well as techniques for their reintroduction and management, a skill set both old and cutting edge.

Who Is This Book For?

Many of my friends are young organic farmers and homesteaders, striving mightily to both detoxify and bring justice to our food system. They are repairing soils, building communities, and feeding the needy. I don't write this book to second-guess them but as an offering. What if those pollinator strips the farmer plants also yield premium wild teas? What if the windbreak contains nutrient-dense native fruits and nuts? What if that marginal swamp or woodland area is the source of high-value medicinal roots? This book is a sourcebook on how to have a more economically as well as ecologically diversified farm or homestead.

This book is also written for my fellow ecological restoration practitioners. We are working to repair millions of acres of degraded landscapes, often with limited budgets and time frames. We can design and implement beautiful restorations yet despair of who will maintain them two or five years from now, let alone decades down the line. We desperately need a human culture to develop around land stewardship. We need humans to be keystone animals that steward and tend diverse wild communities. We need ecoculture, where our economic interests and culture are deeply enmeshed with the health and abundance of the natural landscape.

Perhaps you are a land steward at a nature preserve. What if the surrounding community showed up in droves to remove invasive species, tend rare plants, and disperse seeds and seedlings of desirable species? What if they did this continually, as part of their culture? What if every year, you could host the best potluck ever,

American hazelnuts (*Corylus americana*).

restoration, learning details about growing, placing, propagating, and sharing plant species that exceed what is available in the scientific literature. It is time to expand our vision past supporting "birds, butterflies, and bees" and fully integrate the most challenging animal of all — the human being — into our native plant gardens.

A book on ecology needs to root into the particular, even if it is broadly applicable. Throughout, I tell stories based on my field experience as a botanist in New Jersey, or informed by my travels throughout the Northeast. The closer your region is to my home ecologically speaking, the more of an exact fit you'll find in these stories and in my portrayal of plant communities. Nevertheless, much of what is here is relevant ecologically to eastern North America as a whole, and the ecological restoration approaches and ideas about culture may have a wider scope yet.

I see the path to healthy humans, healthy nature, and healthy societies as deeply entwined with our food and medicine economy. This book is offered to those who feel called to choose natural foods and medicines, native diversity, and cooperative relationships over the toxicity, antagonism, and competitive struggle of an extractive food economy and society. It is written from my experience as a field botanist, native plant grower, and forager, and most of all, from my desire to see wild plants and humans share community again.

gathering your community of stewards to harvest and prepare a sumptuous meal from the abundance of the preserve, with wild meats, fish, fruits, nuts, herbs, shoots, and tubers featured in a deeply flavorful, nutrient-dense, health-promoting feast?

This book is also for gardeners, especially for those in the burgeoning native plant movement. Our gardens are expanding past the typical foundation plantings and ornamental beds, and becoming something else: landscape restorations at the scale of the home or schoolyard. Gardeners can be at the vanguard of ecological

CHAPTER 1

A Different Way

WHEN EUROPEAN COLONISTS first invaded what is now known as the Americas, they may not have had a concept that allowed them to understand the nature that they saw. We *still* don't have a word for the concept in English — a word for the Indigenous practice of honing ecosystems for food abundance while retaining native species diversity and function, though I prefer "ecoculture" to terms like "niche construction" and "engineered environment" used in the academic literature on the subject.

Recent archaeology suggests that populations of Indigenous peoples in the Americas were much higher in pre-Columbian times than previously reported. Current estimates are in the vicinity of 20 million people. By the 17th century when immigration from Europe became widespread, Eurasian diseases had already decimated populations of Indigenous people and lessened the imprint of Indigenous management practices on the land.[2]

Understanding this opens the door for an increasing awareness that Indigenous peoples transformed the landscape in fundamental ways across the Americas, ways that have not been appreciated or noticed by non-Indigenous explorers, scientists, or historians, and were rarely acknowledged by settlers in the Colonial era.

Many of the signature "wild" landscapes of the New World are now understood to be the result of partnerships between humans and other forces. Even the Amazon, that paragon of wildness and biodiversity, may have been in large part shaped by human activity. Likewise, both the tallgrass prairie and much of pre-colonial California were dependent on the cultural activities of people for the composition and type of habitats present. It is increasingly accepted that across the Americas, native peoples shaped the land to create productive systems of what we would now dub "permaculture," "agroforestry," "orcharding," and "game management," but in unique iterations that spawned whole regional ecologies, melding technologies such as fire management, culling, pruning, planting, orcharding, mound and midden building, hydrological manipulation, and the creation of soils.

While recent archaeology is scant in our region, and the land so altered by colonial and modern development that it's hard to read into, accounts written by early European settlers are striking. Early accounts describe extensive parklike woods brimming with nut and fruit trees, and an explosion of wild fish and game. Coupled with evidence from palaeoecology and anthropology, this suggests that, throughout the Americas, Indigenous peoples managed a vast food landscape via fire, plant introductions, and other ecological management techniques.[3] Unlike Eurasian agriculture, these practices didn't depend on the elimination of native species and diversity, or on the active and constant control of domesticated animals. Instead, the entire landscape was managed for abundance, for fruits, nuts, greens, tubers, medicine plants, craft plants, as well as game and fish.

Why is this still breaking news? By the time Europeans established a significant presence here, as much as 90% of Indigenous people may have *already* been wiped out by introduced Eurasian plagues. Once-vast cultural landscapes were already overgrown and untended, and in many places the native peoples and cultures were reduced to a small group of survivors.

In addition, European settlers may have simply lacked the cognitive tools (or interest) to understand an ecologically managed food landscape. Instead, they held an intense prejudice against the Indigenous peoples, branding them as pagans, devil worshippers, and non-humans — a corollary to the usurpation of their territories.

In the Northeast, we rarely consider or recognize Indigenous cultural landscapes. Some are developed over or transformed beyond recognition. We might find groves of pawpaws, bur oaks, hazelnuts, wild plums, or honey locusts, and speculate that they could suggest a former Indigenous village site. Or "out-of-range" populations of other exceptional food plants like groundnut, persimmon, or Jerusalem artichoke. The totality of the system comprising plants, wildlife, and humans, and the ecological knowledge and management techniques that guided it is highly fragmented by colonial-era land appropriation and the aggressive suppression of Indigenous cultural practices.

There is a subset of archaeology known as experimental archaeology. Looking to immerse themselves in the material culturals they are studying, experimental archaeologists craft their own spearpoints and sandals, bowls and baskets, testing the time, effort, and materials necessary. By learning crafts "experimentally," these archaeologists involve themselves experientially in what otherwise might be a realm of abstract speculation.

We too can be like experimental archaeologists, trying materials, seeking techniques, arriving at a thing of beauty and utility. Rather than knapping flint or pressure flaking razor-sharp obsidian, our tools are plants, seeds, soil, fire, and water, as we experimentally replant and care for the once and future food forests of this continent. We're seeking neither the Garden of Eden nor the agriculturalist's yoke, but instead a third way, where everyone is invited to the potluck feast — plants, wildlife, and two-leggeds alike.

Some of the most diverse and productive landscapes on Earth are now understood to be anthropogenic — significantly modified, created, and maintained by people. Does this negate the power and glory of wild nature? No. But it also means that humans can do more than just destroy and defile. We've created sustainable, diverse food systems based around native ecology before, and we can do it again.

Tending the Wild

Remember what I said about the Manhattan of 400 years ago? The book I read (that brought a tear to my eye) was called *Mannahatta*, as was the island we now know as a borough of New York City. In addition to its 30 species of native orchids and old growth forests, Mannahatta featured 55 different ecological community types. According to author Eric Sanderson, Mannahatta had "more ecological communities per acre than Yellowstone, more native plant species per acre than Yosemite, and more birds than the Great Smoky Mountains National Park".[4] It was a stunning example of North American biodiversity.

But this older Manhattan was not an uninhabited wilderness. It was an incredibly diverse island, and it was inhabited by the Lenni Lenape people. Rather than destroy the island's biodiversity, it is likely that they contributed to it through their land management techniques, including the use of controlled fire. Sanderson suggests that "nearly all of the island may have burned on a patchwork, but regular basis." In fact, Harlem was an open grassland at the time of initial Dutch exploration, a fertile meadow of approximately 150 acres over calcium-rich Inwood marble. It was maintained in its open condition through Lenape burning.[5]

How can we reconcile the presence of these people with the extraordinary biodiversity they lived alongside? It is typically expressed that Indigenous people lacked the wherewithal to destroy the habitats they lived in. I think this perspective reveals little about Indigenous lifeways, and rather more about the narrative of "progress" we've all been inculcated with. The need to tend and preserve biodiversity is inherent within Indigenous lifeways such as those practiced by the Lenape. Whereas for most agricultural practices there is a need to replace natural habitats with cropping systems, Indigenous peoples who foraged, hunted, and practiced ecological

Trout with milkweed tops, wood nettle, nodding onion, and bee balm (prepared with David Alexander).

management depended on fully functional ecosystems. For gathering and hunting peoples, *biodiversity equals food diversity.*

Consider the following dietary diversity estimates from a range of Indigenous peoples across the globe:

Dietary diversity: Average number of plant species/cultigens consumed[6]

- Hausa:119
- !Kung: 85
- Tibetan indigenous: 168
- Cherokee: 80
- Contemporary American: 30

Note how lacking in variety our contemporary diet is compared to that of peoples living within a diverse ecology. Consider the effects of that lack of diversity on our nutritional health and internal microbiome, both so important to resisting the chronic diseases that are a major cause of death in the "civilized" world.

What does it look like if we see people in a positive feedback loop with biodiversity, giving as well as taking?

The land management techniques of California's Indigenous cultures natives are described in the book *Tending the Wild* by M. Kat Anderson. She describes how the gloriously abundant California landscapes perceived by Europeans (including nature lovers like John Muir) as untouched wildernesses were in fact the result of different types of Indigenous management. Accordingly, "[t]he productive and diverse landscapes of California were in part the outcome of sophisticated and complex harvesting and management practices." These practices included "coppicing, pruning, harrowing, sowing, weeding, burning, digging, thinning, and selective harvesting." While we tend to think of human-caused disturbances as reducing abundance and biodiversity (when in service to agriculture or civilization building), California natives' techniques maintained the vast wildflower fields containing edible camas bulbs, ancient groves of oaks for harvesting of acorns, and numerous species for basketry, medicine, fiber, and other needs. California had one of the highest densities of hunter-gatherers anywhere on the planet, and they supported themselves and biodiversity in a healthy landscape — one that, by contrast, is now ravaged by wildfire, avalanches, and drought.

While the Northeast surely also featured extensive Indigenous land management, public records of these practices are scant because many eastern Indigenous peoples were killed, displaced, or forced to

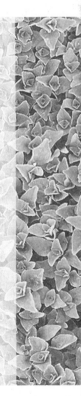

Traditional Ecological Knowledge and This Book

Sometimes land managers formulate goals about how habitats should look and function based on a supposition that the land was "natural" before 1492 and if we can just get it back to how it looked back then we'd be in great shape. On the one hand, sure! I'd love to take a walk through the Musconetcong Gorge or Delaware River Valley in 1491. It would be different, fascinating, revelatory. I'd probably be in tears. But not because I'd be encountering nature in a pure, "wild" state with people deleted from the picture.

A framework that understands the pre-colonial land as "natural" risks writing people out of the equation, in this case Indigenous people who have been written out of the equation in very pernicious ways for several centuries now. It's a form of intellectual dispossession. It also reinforces a narrative that locks all of us humans out, by positing a goal of "naturalness" rather than acknowledging our universal potential and heritage as members of the ecological community.

It's not just 1492, though. Contemporary Indigenous people often get written out of the dominant narrative on "nature" and what to do about it as well. It is another form of dispossession to speak of Indigenous cultures as only entities of an idealized past, with little or no modern relevance.

It is partially to address this that the concept of Traditional Ecological Knowledge (TEK)

has gained increasing recognition within the fields of ecology and land management.

"[Traditional Ecological Knowledge] is a cumulative body of knowledge and beliefs, handed down through generations by cultural transmission, about the relationship of living beings (including humans) with one another and with their environment"[7] writes Fikret Berkes, who studies natural resource management among Indigenous communities. Framing the ecological practices of contemporary Indigenous people as TEK has functioned to politically legitimize Indigenous practices as well as to lend them discursive power in fields dominated by scientific professionals and techniques.

Practically speaking, TEK involves the direct interaction of people with nature through activities such as gathering plants, hunting, fishing, herding, building, healing, and land management. It is not produced by a separate class of scientific professionals but is generated by the community, in the interest of survival, ethical behavior, sharing, and long-term stewardship. Some practices in TEK might have their roots thousands of years ago, but all cultures are incredibly fluid and it is not necessary to assume that an idea is deeply ancient just because it comes from Indigenous peoples, or in order to validate it. It is a combination of long-term residency as well as the direct participation of the community in ecology that gives TEK its practical applicability and social gravitas.

To deny that "[m]uch traditional knowledge has been lost to time and forced assimilation" as Robin Kimmerer and Frank Lake assert in their article about Indigenous prescribed burning (and its suppression) would be a form of historical denial. However, as they go on to write, "much persists in the oral tradition and practices of contemporary native communities, who are only rarely consulted as equal partners in land management."[8] Plant names, place names, ceremonial practices, management techniques, traditional stories, crafts, and all manner of intellectual and spiritual understandings carry Traditional Ecological Knowledge into the future.

As a non-Indigenous person, it may not be for me to know or decide which practices are current, dormant, or have been lost due to genocide, marginalization, dislocation, or just the changing needs of dynamic, living cultures. For this reason, I use the present tense in this book regarding Indigenous cultural practices except where my source explicitly places those practices in the past.

Traditional Ecological Knowledge is a cultural belonging of specific Indigenous nations. In this book, I have decided to convey practices that are already described in the public record, in academic and popular media. I hope I have done so with respect and in the service of the native ecology which is critical to all of us, indigenous to North America and otherwise.

practice their culture in secret well before the birth of any kind of respectful anthropology or ethnobotanical discipline. While the cultural practices which inform Indigenous land management may persist, the opportunity to implement practices such as controlled burning was vastly diminished in much of the Northeast by the 1700s.[9]

Eat Local

Can we extend the idea of eating locally to include eating locally native plant species?

There's a significant movement underway to consume locally produced goods. This is especially true of food. The movement to "eat local" supports local farmers and foodways and helps to disengage from the problems of mass transportation of food.

A great opportunity lies in the consumption of bioregional foods. These are native flora and fauna adapted to the particular place we inhabit. These indigenous foods are knit into the ecology of a place, supporting the vitality of the soil, water, and wild plant and wildlife communities, as well as human needs. These foods support our local ecology in a way that typical farm vegetables and row crops never can.

To truly eat locally is to eat the indigenous diet of a place. Imagine the following spring meal. Our freshly caught trout (probably a stocked trout these days, but once upon a time …) is lightly smoked in hickory as it is cooked on an outdoor fire,

and sprinkled with bee balm and sea salt. As sides, we sauté milkweed shoots in maple syrup, and fry up some groundnut tubers until they are crispy on the outside and soft and starchy on the inside. For a piquant touch, we'll sizzle some morels with wild leek leaves in a skillet until the leaves puff up like fresh tortillas and the mushrooms are browned at the edges.

This is a truly high-end gourmet meal, ranking well in both flavor and nutrition. It epitomizes the indigenous culinary possibilities of our area, using all-native ingredients to create a meal that is so much more than "survival" food.

As much as I'd like to say we can all go out and forage this meal (and we probably should at least once), there may be too many humans in our area for this to be sustainable. So, should we give up and eat hydrolyzed soy protein with corn syrup instead?

I believe we have another option, one of benefit to ourselves and to the wild world. We can restore the degraded habitats that humans created, stock them with indigenous food plants (and thus, teeming wildlife), and create a truly local food system. One that's not based on monocultures of domesticated annual plants from every continent, but from a bioregional palette of deeply delicious, nourishing, and ecologically restorative species.

This book is a guide to the sustaining plants of the Northeast — those that bring us food and medicine. It is a guide to communities — how these plants assemble into mutually supportive groupings in response to specific habitat conditions.

And it is a guide to restoring those communities — with ourselves as members of the community.

CHAPTER 2

Plants in Relationship

I BELIEVE THAT WE CAN practice ecological restoration in a way that explicitly addresses the human animal's need for sustenance and relationships.

Without relationships, individual parts lack meaning. Consider an example from the world of native plants and wildlife. Cardinal flower is a stunning wildflower whose blooms have long red tubular corollas with a sweet nectar reward at their base. Ruby-throated hummingbirds have long tongues and bills that can reach deep into the scarlet flowers to retrieve sustenance. Cardinal flower feeds hummingbird, and hummingbird helps effect reproduction for cardinal flower. Without each other, each one is decontextualized, strange. It is actually the relationship that is emergent and meaningful.

There are many broken relationships in our ecologically devastated world, but perhaps the most broken of all is our relationship with the rest of the natural world. Therefore, a potent ecological restoration

Cardinal flower (*Lobelia cardinalis*).

practice doesn't just repair "nature" — with the human realm edited out — but explicitly focuses on relinking us errant, individualistic, decontextualized animals to each other and to the living Earth of which we are a part.

Plant Planet

Gaia theorists describe the Earth as a biogeochemical entity that creates the conditions for its own persistence, with life itself maintaining the oxygen content of the air, the salinity of ocean water, and the water cycles that seed the clouds with rain.

Plants are the primary converters of the sun's energy into sustenance for all the other terrestrial life on the planet, and important components of the cycles that maintain the atmosphere and the infiltration and transpiration of water. Plants mediate the energy of the sun and make it available to all other life on the planet. Much of the oxygen we breathe comes from plants. They supply fuel, food, shelter, clothing, medicine — all of life's essentials. As animals, it is our doom and our gift to live off the work of others, consuming plants, or herbivores that process plants for us. Like all gifts, it is one we have an opportunity to reciprocate.

Because plants are so important to planetary flows of energy, water and gasses, as well as to our basic needs as animals, many of the problems civilized humans have created can be addressed with plants.

Carbon and Soils

Imagine you are standing in a tall grassland in late summer. Indian grass is head high — you are face-to-face with its seedheads. Run your hand up the stem and your palm fills with golden seeds as you strip the narrow panicle, spikelets shedding easily, seeds begging to be dispersed. Interspersed among the tufted grasses, you see the candelabra blooms of showy goldenrod, the intense purple disks of late purple aster, and the already dry seedheads of wild bergamot, with their thyme-like aroma volatilizing in the heat. Migrating monarch butterflies are nectaring at the asters while a throng of bumblebees visits the goldenrod's yellow spikes.

A beautiful place, and one deeply important to solving the problems of our time.

The roots of Indian grass reach at least ten feet into the earth. Grasslands are extremely productive of soils, and some prairies had 100 feet of topsoil before being destroyed by John Deere's plow.

A single acre of intact grasslands can sequester two to five tons of carbon per year. Plants take what is one of the biggest problems of our time — excess atmospheric carbon — *and turn it into sugar.*

Plants can be likened to carbon straws that pull gasses out of the atmosphere and into the soil, uptake water and minerals, and fuel it all with sunlight.[10]

Sun to sugar.

These sugars, simple and complex, are used to build plant tissues, and to feed the soil biome that supports plant growth. Some of that carbon ends up in plants, some in the digestive tracts of aboveground animals. A lot is exuded into the soil to support bacterial and fungal communities that return the favor by providing water and nutrients to the plants.

Here's some good news. The amount of carbon stored in soils (2,500 billion tons) and plant and animal life (560 billion tons) is nearly *four* times that in the atmosphere (800 billion tons).[11]

So: reducing fossil fuel emissions is critical. We don't want to be adding to the problem. However, the ultimate goal is the re-placement of carbon into terrestrial pools in soils, flora, and fauna.

That's what plants do. They are a 100% solar-powered, pollution-free, renewables-based, honed over millions of years solution to excess atmospheric carbon.

Those Indian grass seeds in your hand? There are many places in this country they could go. Forty-five million acres of turf grasses, shorn for monocultural lawns. One hundred million acres of road medians — about the same amount of space as all the state and national parks combined! We also have over 900 million acres of farmland, largely in annual crops. Much of it conventionally farmed, not very biodiverse, not building soils. Imagine that land restored to native plant communities that sequester

Indian grass (*Sorghastrum nutans*) in flower.

carbon more effectively, while providing food, medicine, and nature connection for people.

Temperature and Water

We are in a deep forest. Red oaks and sugar maples form the canopy, with a few tuliptrees and sassafras that grew up

Mayapples (*Podophyllum peltatum*) after the rain.

minerals and water to the tree roots, extending their range and capabilities.

It is a hot summer day, but beneath the foliage of the oaks and witch hazels the rocks are still cold to the touch and the mosses a verdant green.

Northeastern forests are cool, moist environments. Plants, especially trees, lower temperatures in two ways: by providing shade, and through evapotranspiration.

Shaded habitats can be 20–45°F cooler than the peak temperatures of unshaded places, shielded from the direct heat of the sun's light.

Evapotranspiration can reduce peak summer temperatures by 2–9°F. Plants pull water from deep in soils, and in the course of photosynthesis, they open leaf pores that release water vapor. This vapor can directly form clouds, and it can act in concert with another process, where organic aerosols released by trees and other plants form particles that "seed" clouds. These processes can bring about rain directly, and can also change climate patterns. In the Amazon, the seasonal increase in rain caused by trees shifts wind patterns and brings even more water from the ocean.[12] Some climate scientists suggest that water dynamics and the moderation of temperatures resulting from evapotranspiration are actually far more important than atmospheric carbon in driving climate dynamics.

When rain falls to the ground, it is largely through the agency of plants that it remains

in light gaps. In the understory below, witch hazels thrust loping trunks towards sunny gaps. Black cohosh, bloodroot, rosy sedge, marginal woodfern, and other herbs grow from the moist soil, punctuated with boulders clad in mosses and lichen. Chanterelle mushrooms are borne from mycorrhizae that live in symbiosis with the oak trees. These fungi derive their carbohydrates from fine tree roots while returning in place. First, plant cover shields soils from raindrop impacts which otherwise cause run-off and erosion. Second, organic matter from plants builds soils that have large pore structures. These pores infiltrate and retain water in much the same way as a sponge.

Given organic matter from decaying plants, rain seeded from trees, oxygen from copious pore spaces, and direct sustenance in the form of exudates from plant roots, a diverse soil biology forms that perpetuates the very processes by which it is sustained. Plants have been shown to feed their soil biomes (including the mycosphere, like those chanterelles) with as much as 40% of the carbohydrates they derive from photosynthesis. It is this reciprocal exchange that allows trees to tower a hundred feet over the soil and dozens of feet deep (one report documents a tree's roots extending 174 feet into the soil, though this is clearly exceptional).[13]

We feel this viscerally on a hot summer day, when we walk from a hot parking lot or sunny meadow onto a woodland trail. Beneath the towering oaks and maples, the air is cooler and moister. Leaf litter and coarse woody debris are damp on their undersides. We find salamanders beneath logs and under rocks, and the soil is soft and hydrated beneath our feet.

Wild Plants as Food

What is the characteristic diet of the human animal? Because we inhabit so many of the Earth's ecosystems, and have derived sustenance in many ways, the human diet is characterized by diversity.

Homo sapiens has walked this planet for the last 300,000 years. Perhaps 10,000 years of these have featured agriculture as we know it. Thus, all *Homo sapiens* subsisted on non-domesticated food sources for the majority of our time as a species. Our internal ecology has evolved in the context of consuming non-domesticated species.

During the domestication process, cultigens (domesticated species) often lose nutrient content and phytochemical diversity relative to their wild relatives. Important characteristics of wild plants are removed during the domestication process. Domestication creates larger fruits (but not necessarily with more flavor), less challenging flavors (such as the removal of bitters), softer leafy tissue, and other characteristics for which plants are bred, such as shelf life or easy germination.

Wild plants respond to the rigors of wild living, including herbivory, pathogens, and other environmental stresses, through the production of a diverse palette of phytochemicals. Domesticated plants no longer need to produce adaptive phytochemicals, because many environmental stresses that wild plants face are managed by humans through supplemental watering, weeding, insect control, and soil amendment.

Wild plant species are generally superior to domesticates in their content of pro-vitamin A, vitamin C, minerals such as calcium, phosphorus, iron, and potassium, polyphenols (which help prevent cancer and various degenerative diseases), and omega-3 fatty acids.

Botanist Arthur Haines writes, "[o]ur body's ability to carry out metabolic processes, heal from injury or sickness, and defend itself from pathogens has evolved concurrently with a [wild foods] diet that is very different from the one most people experience today."[14] In addition to containing critical nutrients and vitamins, wild plant foods trigger body processes from digestive enzyme secretion to immune response, and modulate our internal ecology — the bacterial symbionts that live within us. Now we suffer from everything from digestive issues to depression because our internal ecology is disconnected from the external ecology in which we evolved. A modern diet of simple starches and muscle meat may not suffice in supplying our need for biological complexity in our diet.

Wild Plants as Medicine

I appreciate the way that herbalists think. They consider the ecology of the human body. Their healing practices utilize concepts that I think are useful for ecological restoration.

Herbalists often contrast their practice with conventional medicine — the dominant medical model. Conventional medicine must be credited for its efficacy in trauma medicine — keeping people alive in the midst of an acute crisis like a heart attack or the aftermath of an auto accident. If I have a shattered bone sticking out of my flesh, by all means bring me to the hospital.

Conventional medicine is good at acute crises, but often fails to support the continued health or recovery of fundamentally healthy people. It offers little for the *prevention* of the chronic illnesses that are leading causes of death and suffering in much of the world. It is not so good at

Giant Solomon's seal shoots and wild leek foliage.

what I would consider healing: prompting and supporting the body's numerous mechanisms for self-repair and recovery.

Herbalists seek to use plants to modulate our inner ecology, support the body's ability to heal itself, and trigger enzymes, hormones, and systemic responses. Part of the genius of plant medicine is that it acknowledges our fundamental kinship with plants. Plants exist in the same terrestrial habitats we do, and need to respond to many of the same challenges. Because they have neither animal mobility nor mammalian immune systems, plants have addressed many environmental challenges through the production of phytochemicals. These challenges include bacteria, viruses, fungal attack, heat, cold, and oxidative stress. Plants offer medicines that serve us directly in dealing with these kinds of environmental stresses. And because they've had 400 million years or so to evolve these defense mechanisms, the solutions arrived at by plants can be more sophisticated than pharmaceuticals created in a laboratory setting.

Contrast, for example, the way that pharmaceuticals and some plants deal with bacteria. There are many species of potentially pathogenic bacteria that reside in our bodies as elements of our internal biome, including species such as *Staphylococcus aureus, Pseudomonas aeruginosa, Streptococcus* spp., and *Escherichia coli.* These are commonly thought of as disease-causing "germs." However, these species only become problematic when our internal ecosystem is so disrupted that they can begin to multiply, proliferate, and overwhelm our body systems.

Conventional medicine prescribes antibiotics for these bacteria. Antibiotics are chemical agents meant to kill bacteria. However, antibiotics are problematic because they also kill a lot of other species in our internal biome, species that are crucial for processes like digestion, and killing pathogenic species tends to breed for resistant bacteria.[15]

By contrast, an interesting class of plant chemicals takes a different approach to suppressing pathogenic bacteria, by inhibiting quorum sensing.

What makes bacteria so dangerous is their ability to form larger entities such as biofilms and plaques where a multitude of bacteria "team up" and dominate infected areas of the body to the detriment of bodily function. In order for bacteria to team up and become pathogenic, they utilize a system of communication known as quorum sensing. Essentially, this is like bacteria asking, "Do I have a crew?" It is a precursor to the formation of biofilms and plaques.

Rather than rely exclusively on antibiotics, the plant kingdom has developed a number of chemicals that inhibit quorum sensing. Phytochemicals including quercetin, sinensetin, apigenin, and naringenin display anti-biofilm formation activity against bacteria.

Plant tinctures.

This is an elegant solution that does not foster resistance the way that antibiotics do. Phytochemicals with anti-quorum sensing activity merely inhibit the ability of potentially dangerous bacteria to enter into a disease-causing state, thus rendering them virtually harmless.

Not all herbalists view plants in the strictly functionalist terms I describe above. Many consider the spiritual, ceremonial, and emotional powers of plants, and the cultural practices associated with them, to be of equal or greater importance. Certainly this is true of many Indigenous medicine practices, and this way of interacting with plants deserves acknowledgment and respect.

Ultimately, herbalists are healing internal terrain using plants, and restoration practitioners are healing external terrain using plants. For ecological restorationists, herbal medicine can serve as a metaphor and an outside reference point. It is a mirror for what we might strive to do in the best ecological restorations, which is to support an ecosystem's ability to heal itself, possibly even with some plant medicine of our own. We benefit from cross-disciplinary conversations.

Plant Communities

Riparian Forest

Ostrich fern (*Matteuccia struthiopteris*) along the Delaware River.

Catskills stream.

Pawpaw (*Asimina triloba*) along the Susquehanna River.

Purplestem angelica (*Angelica atropurpurea*).

Mesic Forest

Above: American ginseng (*Panax quinquefolius*) is found in mesic soils on rich geology.

Right: Rich mesic forest indicators — maidenhair fern (*Adiantum pedatum*) and wild ginger (*Asarum canadense*).

Left: Black cohosh (*Actaea racemosa*) flowering in a light gap in the forest.

Right: Spicebush (*Lindera benzoin*) in fruit.

Upland Oak Forest

Hilltop with upland oak species and heaths.

Downy serviceberry (*Amelanchier arborea*) in bloom.

Smooth Solomon's seal (*Polygonatum biflorum* var. *biflorum*) freshly emerged in spring.

Chestnut oak forest with an understory of black huckleberry (*Gaylussacia baccata*).

Glades

Nutrient-poor glade with pitch pine (*Pinus rigida*).

Left: Black huckleberry (*Gaylussacia baccata*) is typical of nutrient-poor glades.

Right: Dittany (*Cunila origanoides*) is often found in the vicinity of glades on nutrient-rich geology.

This glacially scoured ridgeline hosts a wide diversity of dwarf fruiting shrubs.

Meadows

The meadow at our farm was a soybean field eight years ago.

Wild bergamot (*Monarda fistulosa*) seeds in readily to created meadows.

Common milkweed (*Asclepias syriaca*).

Blackcap raspberry (*Rubus occidentalis*). *Rubus* species will often occupy transitional zones between meadows and forests, both spatially and temporally.

Swamp in springtime with marsh marigold (*Caltha palustris*) in bloom.

Highbush blueberry (*Vaccinium corymbosum*) in forested wetland.

Mad dog skullcap (*Scutellaria lateriflora*) in flower.

Goldthread (*Coptis trifolia*) on the margins of a conifer-dominated swamp.

Sunny Wetlands

Kerry Barringer kayaking in a beaver dammed marsh at one of our botanical survey sites.

Black elderberry (*Sambucus canadensis*) on the shore of a pond.

Left: Broadleaf cattail (*Typha latifolia*) with pollen-laden male flowers.

Below: White turtlehead (*Chelone glabra*) is adapted for bumblebee pollination.

Northern Forests

Boreal forest lakeshore in Maine.

Rachel walking through a groundcover of wild sarsaparilla (*Aralia nudicaulis*).

Purple-flowering raspberry (*Rubus odoratus*) on a high bank in the Catskills.

Yellow birch (*Betula alleghaniensis*) with exfoliating bark.

Sandy Pine Forests

The Pine Barrens.

Pitch pines (*Pinus rigida*) are adapted for the frequent fires typical of sandy pinelands.

Highbush blueberry (*Vaccinium corymbosum*) fruits abundantly in sunny openings.

Wintergreen (*Gaultheria procumbens*) is a diminutive subshrub which thrives in nutrient-poor soils.

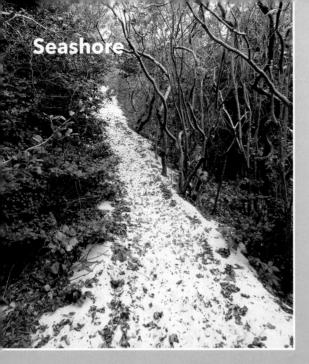

Maritime forest with Canada serviceberry (*Amelanchier canadensis*) and American holly (*Ilex opaca*).

Sea rocket (*Cakile edentula*) in front of the foredune.

Above: Prickly pear cactus (*Opuntia humifusa*) on the hot sands.

Right: Beach plum (*Prunus maritima*) with sun-ripened fruit.

CHAPTER 3

Ecological Restoration

AMIDST ALL THE GRIM NEWS of ecological collapse, we have an opportunity. It is an opportunity to re-imagine our place in nature, and use our artful, technical, and mimetic capacities to create places that emplace us.

Our practice is ecological restoration, our tools are both scientific and cultural, our inspiration comes from remnant natural places, and the building blocks we work with are wild plants.

Wild plants are the link between the realms of sun and sky and those of soil and stone. They are the photosynthetic beings that channel solar power and energize the relationships from which ecology emerges. They are the fundamental tools of the human healing arts, from the medicine person to the restoration practitioner.

Mimesis and the Reference Ecosystem

I want to talk to you about mimesis — our ability to engage the magic of copying, miming, and replicating that which we observe.

Scientists might describe mimesis as a function of mirror neurons. Cultural anthropologists might see in mimesis the deep roots of drama, story, empathy, and intersubjectivity.

When we observe the movements of another being, mirror neurons fire in our bodies, reenacting what we're witnessing, laying down patterns that we can repeat. The more nuance we feel our own bodies with, the more nuance with which we can mirror the movements of others. These "others" are not limited to human others. We can watch a big cat like a tiger, or a prey animal like an antelope. Stretch like one, leap like another — even if just in our mind.

With deep observation, our bodies can anticipate the next movement of those other than us. We can watch a dancer or a martial artist and know what they will do next, at least when we share their practice and have laid down the same paths in our own mind. In this way we can "become" other creatures.

Honing mimesis to a fine art, trackers can embody the movements of their prey, reading their body language in tracks that suggest pauses, hesitations, accelerations, and a full range of other gestures. Embodying each of these, our human ancestors (and some of the world's few remaining hunter-gatherers) can literally read the minds of animals. By becoming them.

Mimesis is basic to human learning. Movement practitioner Simon Thakur suggests that the earliest stories of our prelingual ancestors may have been dramatic pantomimes, perhaps around a fire or on the trail.[16] To this day, observational learning plays a significant part in our social development and movement practices, though in other fields it has been diminished in favor of a model of learning in which the mind is made to memorize and drill data.

At the root of ecological restoration is a mimetic practice. Fundamental to restoration is the concept of the reference ecosystem: a natural area with exemplary diversity and integrity. By visiting reference ecosystems, we can immerse ourselves in an observational mode of learning, returning not just with data but also intuitions about structural patterns and relationships. By visiting an exemplary natural area, we have an opportunity to witness the relational aspects of ecology (not just atomized parts) by perceiving a real ecosystem in which each entity bears myriad relationships to another. Too complex for writing down or depicting in data form, our mimetic capacity is engaged instead. We immerse ourselves in a reference ecosystem and then seek to embody it in our restoration site.

The translation from reference to restoration mirrors in some ways the fundamental human capacity for storytelling, re-enactment, drama, and embodiment of the other — one of our deepest ancestral skills and defining qualities. Ecological restoration melds science with the mimetic art at the root of being human, awakening connectedness as we *become* nature in order to heal ecosystems.

Restoring Plants

A five-foot-tall, decades-old perennial wild plant like black cohosh doesn't just "appear" out of the ground in the spring. It punches out, purple and unfurling, both graceful and primeval. From atop its thick, dark stem, three separate appendages unfurl into horizontal plateaus of compound foliage, each deeply incised leaflet spreading out until the leaves collectively occupy several square feet.

If the plant is to flower that year, a stem later shoots up over the primary foliage. The flowering stem briefly holds a miniature cluster of flower buds, like baby corn, above reduced lateral leaves. In midsummer, the flowers open, small whitish starbursts of stamens on an upright stemmed stalk called a raceme. Blooms open sequentially from the bottom of the raceme to the top, an upwards-moving column of white taking several weeks to reach from the bottom of the raceme to the narrowed tip.

There the flowers stand, chest-high or taller above the ground, catching the afternoon sun as it slants below the canopy. A typical colony of hundreds or even thousands of black cohosh plants illuminates the understory, a legion of tall white flowering wands glowing against the dark trunks and long shadows of the forest.

The flowers have a deeply musky odor, a stale smell that enters the nostrils and sits in the back of the nose, like old popcorn smell in a movie theater. Attracting bumblebees, honeybees, and flesh flies, the flowers are pollinated and each fertilized embryo forms a crescent-shaped seed. The seeds nestle in two-lipped capsules, attached by short stems onto the central stalk of the raceme just as the flowers were. As the seeds mature, the entire seedhead turns a deep black and awaits dispersal.

There is a large colony of black cohosh on a flat ridgetop near where I used to live in central New Jersey. Walk there in the late fall, and the dried leaf litter of black oaks crunches beneath your feet. The thin rattle of black cohosh seeds sounds almost subaudibly as you brush past decumbent stems bringing seedpods nearer the earth.

Black cohosh (*Actaea racemosa*).

I once walked a circle around the colony there and found it covered over four acres. The colony on this oaky ridgetop numbers in the thousands of stems, many of which are evenly spaced, though the colony meanders, grows ragged, misses spots, and thrives densely in others.

Once I was explaining to my wife Rachel something I had read — that black cohosh seeds had no documented dispersal methods other than falling to the ground and hoping for the best. This, in contrast to the many species that are specially adapted for dispersal by wind, birds, adhesion to fur, ants, and so on. By all accounts its dispersal distance is limited to the immediate surroundings of the parent plant except in exceptional circumstances. One scientific account I read described its effective dispersal distance as "0 meters" per year.

When I told this last figure to Rachel she disagreed. Rachel pointed to the dried, bleached flower stems left over from this year's cohosh bloom. They no longer stood shoulder height to us, but were tipped over and hovering practically horizontal to the ground — scattering the seeds still held in their rattlepod heads at a fairly regular distance of three to four feet from the parent plant.

Regardless, cohosh is not a fast colonizer by seed. Black cohosh is dispersed neither by bird, mammal, reptile, nor wind, except in rare and accidental circumstances.

Perhaps it is this highly regularized dispersal mechanism — measured faithfully every year in cohosh stems, like some archaic measurement unit of rods or fathoms — that is to account for the regular, and highly colonial distribution of individuals within black cohosh populations.

So how long does it take for a four-acre grove of cohosh to arise, presuming a single parent, spreading largely by seed, one to two yards every year or two, tacking in some extra time for these long-lived and large-rooted forest herbs to reach sexual maturity? The math is beyond me, but it takes long enough, I'd presume, to qualify the plants as ancient. Even if the trees aren't old growth, the black cohosh may be.

Black cohosh must migrate long distances on occasion. The patch on the oaky ridgetop got there somehow. Some seed-laden mud on the hoof of a deer? A hurricane? Regardless, cohosh doesn't easily cross highways, golf courses, or suburban developments. Though we have a higher percentage of forest in the Northeast now than for most of the last two centuries, the majority are young woods, gone feral from abandoned agriculture. Some are suitable habitat for cohosh, but it may be centuries before these post-agricultural woods are actually reached by the small, black, half-moon-shaped seeds falling from their chest-high perch.

Black cohosh has proved a tremendous herbal ally for people. It is one of the best-selling medicinal herbs on the global market, used to balance hormones during menopause, and for musculoskeletal support. In 2017, $32 million dollars worth of black cohosh was sold just in mainstream outlets in the United States.[17] It is in danger of overharvest in the Appalachian forests it calls home. But the problem can lead us to the solution, and towards being better stewards of the species.

Black cohosh has helped us, and we can return the favor by bringing cohosh, plant by plant, seed by seed, to the young hillsides and rocky slopes where it once bloomed and can do so again.

There's a special relationship that we humans have to plants. We can promulgate plants. We can reproduce plants. Other animal species disperse seeds, for sure, but we humans can bring plants to new places in very deliberate ways. With plants, we can take the foundational building blocks of ecosystems, (pro)create them, and make more of them.

Right now, we have many degraded human landscapes that detract from the function of the living Earth. But if our human landscapes are rich and abundant with plant life, then our terrain is functional. Because plants are the basic building blocks for addressing our environmental concerns. Sequestering all of that carbon and maintaining atmospheric gasses in livable proportions. Building living soils. Plants link the world of soil and stone to the world of sun and sky. We need plants because they mediate all of the foundational processes of the current epoch of life.

CHAPTER 4

In Community with Nature

THE MOST POPULAR DEFINITION of a native plant is a species that resided on this continent prior to 1492, i.e., prior to European invasion of the Americas. However, this definition doesn't reveal much about why this distinction is important.

A more functional definition would be that *native plants are those species that have evolved together with the wildlife, geology, soils, and climate of an ecological region.*

Note "ecological region"; political boundaries are moot here.

This definition is fundamentally relational. Native plants have evolved in a close relationship with the climate and geology of our region. They thrive in tandem with the wild creatures that utilize them for food and shelter, creatures that often return the favor by providing seed dispersal or pollination services.

The importance of native species is related to their ecological tolerances, interdependencies, and co-evolution with other biological and geomorphological entities and forces.

Such relationships in our local ecology are legion. Some are well-documented: Trees and mycorrhizal fungi. Monarch caterpillars and milkweed. Oaks and their rodent and avian dispersers. Many relationships are not well-documented. Our natural sciences are still emerging from a mechanistic understanding of nature in which interrogating and taxonomizing fragments such as species and chemical constituents is considered more important than understanding relationships.

Native plants support the food web in ways that exotic species cannot. Fully 80% of invertebrates are specialists on certain lineages of native plants. Invertebrates form the base of the animal foodweb that supports humans and other wildlife. Given precipitous drops in global insect populations in recent decades (a current estimate is a 45% decline), we need to turn serious attention towards restoring diverse native plant communities if we hope to reverse this trend.

Native cherries support almost 400 lepidopteran (butterfly and moth) species. Highbush blueberry, a wonderful food plant for us, supports more than 250 species, and purple-flowering raspberry, a showy and edible bramble, provides sustenance for the caterpillars of over 150 native lepidopterans.[18]

Let's pay at least as much attention to relationships as to parts. Cardinal flower feeds hummingbird, and hummingbird helps cardinal flower reproduce.

Communities

Native plants are parts of biological communities. As with life itself, communities are ever-shifting. Perhaps they are better understood as a flow than as a static entity. Regardless, the important defining feature of a community is that it is characterized by myriad relationships, complex, interdependent, and multidirectional.

People are often characterized as being part of a community — a community of humans, at least. Much talk and effort goes into creating communities, these semi-voluntary associations of people, bonded tightly. What sometimes goes unremarked is that community arises from interdependency, sometimes dire, inextricable, even unpleasant interdependency.

Corporate-structured society has diminished human community by reducing

Spicebush swallowtail caterpillar on spicebush leaf.

Baltimore checkerspot.

our lateral interdependencies, from person to person. There are rich networks of need and exchange when one gets vegetables from the local farmer, healing from the herbalist, grain from the miller, music from the musicians, and so on. These lateral networks have largely been replaced by vertical networks, with all of us at the bottom and a surprisingly small group of corporations and state entities at the top. We go to the corporate distribution centers for our pressing needs, whether for food, medicine, entertainment, clothing, or otherwise.

As much as we're forgetting what it means to engage in human communities, we are even more deeply sundered from natural communities.

What would it mean to be in "community" with our ecological region? We would have more lateral interdependency, less vertical dependency. More food, medicine, entertainment, clothing and otherwise, from plants and animals we personally know and encounter. Fewer staples of life processed and mediated top-down by corporations and the state, less received from the supermarket, the hospital, the screen.

Yet being part of a community entails more than just taking. There is a reciprocal nature to being emplaced. Hummingbird pollinates cardinal flower, cardinal flower feeds hummingbird.

Belonging to an Ecological Community

What would it mean for those of us who are not Indigenous to develop belonging with our ecological community?

Like many other residents of North America, my ancestry lies elsewhere, in Europe. My ancestors were not early colonizers of America after 1492. They arrived after much of the bloody work of genocide and culture war against Indigenous people was considered to be history, though it has continued perniciously. Both sides of my family left inhospitable circumstances in their European homelands. Nevertheless, they benefitted from the bounty made available here by the intense disturbance wrought by earlier European invaders. They came into a land in which riches flowed, as the natural wealth created by others, non-human and human, was made available for exploitation.

What would it mean for us — the North American immigrants, willing and not — to belong to its ecology?

Like hummingbird, we can pollinate. Like blue jay and squirrel, we can disperse seeds and nuts. This is a beginning.

Yet more is needed from us. We arrived as invaders, or in their wake, to a place of vast disturbance. Many relationships in the ecological community have been destroyed. Our role is clear: to restore the ecological community. To abet relationships, engage in interspecies reciprocity, and re-engage in mutualisms and common cause with other species.

To belong to the ecological community, we must become ecological restorationists.

We will not be restoring nature to "virginal," untouched, static perfection. Rather, we need to restore ourselves as human animals, and our citizenship in the community of wild beings. We need to bridge the philosophical abyss between human and natural by healing degraded places and by building communities that include the biotic and abiotic beings of our ecological region. We need to experience our interdependency with the wild world in a way that is less mediated by processing, digital screens, or even books.

We need to cure alienation, heal the earth, and restore relationships to both our human communities and to our interaction with the rest of the natural world. We need to learn from and respect the practices of Indigenous people who have been doing this for a lot longer than settler culture has been here.

One of our fundamental human gifts is a profound ability to observe, imitate, and create anew. This mimetic skill underlies our practices from art and music to language, magic, and science.

In order to become healers of the natural world, we must first become keen observers, immersed, using our animal senses. What do we witness of the complex ways in which habitats assemble?

The old Eurasian agricultural lifeway sought to clear, level, and make homogenous. Our architectural and landscaping practices have been similar. The new ways will work with existing habitats, at whatever stage they are at. If you inhabit a place that has cleavers and mayapple, perhaps you have a site for wood nettle and wild leeks. If you see black cohosh and spicebush, maybe you can support goldenseal and ginseng as well. If you have a lawn, you also have a wild bergamot, common milkweed, bayberry, and wild strawberry polyculture just waiting to happen.

In the next section, we'll learn to read a landscape to understand both its potential and the sources of degradation that may be inhibiting its full and bountiful expression.

SECTION TWO

Learning Your Land

A FEW YEARS AGO, we[19] were commissioned to design and implement a restoration for the East Woods, four woodland acres at Great Road Farm, an organic vegetable farm in central New Jersey. The owner's goals included improving wildlife habitat, and having a beautiful natural area adjoining his home. When we described our plan for a restoration that included many edible plant species that could supplement the farm's production, the owner was intrigued and asked us to proceed.

Our first step was to assess the site, asking questions that I will elaborate on throughout this section of the book. These fundamental queries will serve you as you seek to better understand your land, its restoration potential, and the plants that may be best adapted to its conditions.

In the section below, I'll break down these inquiries into five categories: soils, water, land use, natural disturbances, and indicator plants. As you'll see in my description of the Great Road Farm site, these are all interconnected and observations about one will quickly lead you to ideas about another.

The East Woods restoration site had once been a farm field. Deemed marginal, it was planted to conifers in the mid-20th century. These were never maintained as intended, and a majority of the spruces and pines blew down in Superstorm Sandy in 2012. The site was stark — blown-down conifers strewn about, with the remaining standing trees looking wispy and vulnerable. Beneath the mess was a weedy ground layer hanging on to erosive, damaged soils.

Other challenges to restoration included a legacy of heavy deer-browsing pressure, and the presence of invasive plant species including Japanese stiltgrass, mugwort, and creeping ("Canada") thistle.

The site's geology is red shale, a sedimentary rock that often weathers to form clay soils. These soils can become very waterlogged, especially when they are compacted by agricultural use. At the bottom of a shallow slope there was an area of inundated soil with a small grove of red maples underlain by sedges, good indicator species for wet soils, and beyond that, a weedy strip of the invasive reed canary grass, which also favors wet soils.

We didn't consider the site's wetness to be a negative characteristic, but we did want to reduce the potential for erosion on the slope, and increase the chances that a stable, perennial plant community could take over from the weedier elements of the site's flora. To this end, we used downed logs in tandem with earth-moving equipment to create a series of swales along the slope to capture and infiltrate water, and to serve as temporary pools in the wettest portions of the year.

When we subsequently seeded the site, it was exciting to see various wetland natives, including cardinal flower, Joe Pye weed, and boneset recruit in the swales we had created. These species confirmed our intuition about those areas, born from

Before.

Before and after, Great Road Farm restoration.

After.

a consideration of soils, water flows, and land use history.

With every site, one is faced with a consideration I call "tend or mend." Already flourishing areas may be ideal sites for subtle improvements — tending. With the East Woods at Great Road Farm, we chose to mend — fixing water flows, adding species via seed and direct planting, pulling and smothering invasive plants, and even fencing against deer browsing.

Over a half decade later, the site is booming with native herbaceous plants introduced as part of a seed mix. The many fruit- and nut-bearing shrubs we planted are coming into their own, including American hazelnut, elderberry, chokeberry, highbush cranberry, pawpaw, spicebush, and nannyberry.

You can use similar processes to learn about your own land or restoration site.

Great Road Farm east woods restoration 2014–2017.

CHAPTER 5

Reading Geology, Soils, and Water

Soils and Geology

SOILS ARE LIVING COMMUNITIES comprised of both living and non-living (biotic and abiotic) elements. Soils are influenced by the parent rock they are derived from, as well as by organic matter and living soil organisms.

Parent rock materials in soils include weathered bedrock, and may also include transported rock materials, moved by wind, water, and glaciers. Because rocks contain differing ratios of minerals, the resulting soils derived from different parent rocks may be more or less rich in fundamental plant nutrients such as calcium, magnesium, iron, and potassium.

Different plant species have evolved to thrive with different levels of soil nutrients and different types of soil structure. A soil that is poor for wild ginger may be just right for black huckleberry.

Nutrient levels and pH in soils can also be altered by the plant species living on those soils. For example, basswood,

elm, ash, and some other tree species have a relatively nutrient-rich leaf litter.[20] Conversely, heaths such as huckleberry and blueberry leave a litter that is low in available nitrogen and other nutrients, perpetuating poor soil conditions for many other species (but ideal for their own species). The level of organic matter and the abundance of soil life can have radical effects on the type of plant community that will thrive on a site, and all factors are in interdependent feedback loops.

The intent of this book is to help you match plant species to pre-existing habitat conditions, using appropriate species that are well-adapted for the fundamental underlying conditions of a site. This is in contrast to attempting a radical (and probably short-lived) alteration of soils to match a limited suite of plants, as is often practiced in agriculture and gardening. This does not preclude the restoration of degraded or compacted soils. Each site will have its own balance between tending the existing site conditions, and mending them when ecological repair is imperative.

Water

Fresh clean water is as important a resource as any other. Yet water is often treated like garbage, drained from

Rock Types Influence Soil Characteristics

Igneous rocks are formed from lava or magma and can be classified as mafic or felsic. Mafic igneous rocks form at high temperatures, are most stable at those temperatures, and contain relatively high amounts of minerals such as magnesium, calcium, and iron. These rocks contribute to soils that, for our region, are fairly high in pH, sometimes termed circumneutral or rich soils. These rocks are often dark in color, and include basalt and diabase. Felsic igneous rocks are high in silica (in the form of quartz), and also have increased levels of sodium, aluminum, and potassium. These produce acidic soils that can be poor in terms of crucial plant nutrients. Felsic rocks are typified by granite and related rock types.

Sedimentary rocks are derived from weathered rock materials (sand, silt, and gravel particles) and often organic materials precipitated from solution in water. When sediments are derived from or include crustacean shells (such as sediments formed on lake- and sea-bottoms), the resulting rocks can be calcareous (high in calcium), as in limestone, dolomite, marble, and calcareous

shales. These, too, can form rich soils and have a circumneutral pH. Other sedimentary rocks can be quite poor in plant macronutrients.

Metamorphic rocks result from the action of heat (and sometimes water) on any type of rock. These, too, can contribute to rich or poor, circumneutral or highly acidic soils. In our area, metamorphic rocks such as amphibolite contain rich-soil plant communities, whereas many areas of gneiss can be quite acidic.

Different rock types also weather to form different-sized particles that exhibit different characteristics in terms of water retention, air circulation, and nutrient retention or leaching. For example, sand is composed of large particles that drain readily, leach nutrients easily, and can be quite well aerated. On the other end of the spectrum, clay is formed by tiny particles that possess a negative charge and can attract and bind positively charged ions of desirable nutrients. Thus, while clays may be undesirable due to poor drainage or aeration, they can host rich plant communities that thrive with high nutrient and (sometimes) high moisture levels.

Fringed gentian (*Gentianopsis crinita*) thriving on calcareous substrate.

developed landscapes and agriculture in dirty bursts, incising streams, flooding rivers, dumped to the ocean with a load of chemicals and a flotsam of trash.

Human activities have led to decreased recharge of underground water. This affects not only the water table, but also the vadose zone — below the surface, above the water table. Plants draw water from this zone.

Soil properties affect infiltration to the vadose zone and the water table. Soils with coarse textures and large pores recharge groundwater more readily. If your site has sandy or gravelly soils, chances are that they are very well drained. Organic matter in the soil, and healthy soil biology, improves soil structure and can also allow for water infiltration into soils. That said, if you have a lens of heavy clay beneath your topsoil, it might be better to work with a wetlands plant community than to try to remedy it by extensive work aimed towards drainage. The slow infiltration and filtration capabilities wetlands possess are a virtue unto themselves. Infiltration is also affected by topography, with swales and pit-and-mound topography offering depression storage that slowly percolates water into the ground, even on otherwise steep slopes.

Devegetation, disturbance, grading, ditching, and soil compaction can lead to increased flooding and erosion, and less infiltration into soils. Agriculture has been a significant factor in the exhaustion of groundwater resources and destruction of soils. The flattened, compacted, devegetated winter farm field stands as a stark example of exactly what not to do if we want to retain soils and recharge groundwater.

Perennial plant covers shield soils from raindrop impacts, slowing water before it reaches the soil, allowing rainwater to infiltrate into the soil. By contrast, rain on bare soils causes erosion and migrates fine soil particles into soil pores, clogging them. When such soils are dried out, they can become hydrophobic (repelling water) and difficult to re-wet.

To retain and recharge groundwater, we can reestablish plant communities with diverse structures above ground and also below ground in the root zone. If your landscape has been graded and drained by agriculture or development, the restoration of a naturalistic topography with swales or depressions can also allow for localized water storage and recharge.

A site's geology will also affect its ability to recharge groundwater, act as an aquifer, and manifest surface water. The type of rock present on your site will offer differing degrees of porosity and permeability.[21] For example, a site with massive bedrock with very few fissures will not drain well, whereas a deep layer of glacial till or river deposits, with gravel, cobbles, and boulders, is a very porous substrate and usually will drain readily.

Plants and Hydrology

Inundated soils present a challenge to many plant species, as do excessively drained soils. In the middle is a soil condition we term mesic — moist but well-drained.

Stormwater runoff from corn field.

Many plants thrive in a mesic soil — it's what we aim for when we create a vegetable garden or amended flowerbed. In the wild, *mesic* soils can host diverse plant communities. When richer mesic soils are disturbed, a riot of weeds can ensue, as conditions are favorable to generalists.

Many plant species have found specializing in the extremes of wet (hydric) or dry (xeric) soils to be a successful strategy. Specialists in dry soils — prickly pear cactus, or rattlesnake weed, for example — are often low to the ground, which conserves moisture by minimizing transpiration due to wind convection. Put them in a mesic forest full of black cohosh and spicebush, and they will be rapidly shaded out, though they might relish the moist, well-drained soil. The xeric species are stress tolerators, but not very effective competitors.

Rattlesnake weed (*Hieracium venosum*) basal leaves.

In wetlands, one finds species tolerant of oxygen-depleted, waterlogged conditions. Blue flag iris, swamp milkweed, and cardinal flower thrive in areas where groundwater is at the surface. Yet they will also thrive in a prepared garden and are not dependent on inundation to survive. The aboveground forms, seed dispersal mechanisms, and root structures of these and other wetland species are uniquely adapted to sodden conditions under which neither a xeric species such as prickly pear cactus nor a mesic species such as black cohosh could survive.

CHAPTER 6

Land Use History

WHEN RACHEL AND I first moved out to the woods, I knew nothing about natural areas. I grew up a city kid and my idea of a hike was walking the five miles or so down Broadway to Chinatown to get roast pork buns and comic books.

I was surprised and impressed when a red-tailed hawk came screeching out of the trees and skimmed the roof of the dank little cottage we rented in the Sourland Mountains of central New Jersey. Also when I saw a giant pileated woodpecker calling from a tree, sounding like some tropical monkey garbed in black and red and white.

Rachel and I used to walk up the long narrow dirt road, starting at the driveway across from our moldy little cottage, passing our landlord's pond, then the unfinished house of the crazy Hungarian whose leaking septic smelled just like raw sewage in the Danube in my grandparent's village of Szentendre. We'd cross a few little dirt driveways and a modernish

wood home, then through the domain of Woodpeckerman. We called him Woodpeckerman because there were always woodpeckers near his driveway and woods (and lots of abandoned Volvos too). Red-bellied woodpeckers, downy woodpeckers, pileated woodpeckers — they were always there, drumming and making contact calls.

We got to know our plants, one at a time, and started paying closer attention to those we saw on our walks. We'd walk up the driveway, turn onto the dirt lane, and cross past the old farm pond being swallowed by woods. The woods by the pond were full of multiflora rose, the invasive shrub, and otherwise dominated by short red maples. It was a good spot for tracking deer but not exactly bursting with plant diversity. Having just learned about invasive species, we'd search diligently for native shrubs and trees. Usually the ones we found were bonsaied by deer browsing and half-hidden beneath the armed canes of the non-native rose.

Further down the lane, we'd get to Woodpeckerman's woods, a mesic flat with white oak and black oak, mapleleaf viburnum and lowbush blueberry, rue anemone and bloodroot along the road's margin. And lots of woodpeckers. And no multiflora rose.

This seemingly unfair progression from our thorny thicket to his thriving oak woods remained a mystery for quite some time. Eventually I realized that it had to do with the influence of land use history.

Influences on the composition of natural areas include annual precipitation, soils, bedrock geology, extremes of temperature, and the influence of the biota — from predators to pollinators. Those factors are certainly key, but each of those variables was essentially identical between "our" woods and Woodpeckerman's place. The difference was land use history — the effects of human use during the post-colonial era.

The area around the pond, now full of multiflora rose and young red maples, had been a cow pasture until at least the 1940s. The pasture eliminated the original native forest, changed the soils, and favored intentionally introduced pasture grasses, or naturalized species that happened to be less favored by cattle in an intensive grazing scenario. When cattle were no longer raised here and the dairy farm was abandoned, the former pasture went feral with whatever bits and pieces of flora were

Orchard grass field margin.

nearby. The burgeoning deer population also exerted an influence. As the red maples grew and exerted more shade, shade-tolerant native species might have recruited but many were browsed out of existence by deer. Meanwhile, deer-resistant invasives like multiflora rose proliferated and came to dominate the shrub layer.

Woodpeckerman's oak woods were of older vintage. The oldest maps show this as forested even in the 1880s — not much later than the peak of land clearing in that area. Probably the forest was harvested multiple times for its wood. While trees came and went and canopy cover fluctuated, the native soils, seed bank, and possibly even ground layer species may have remained much the same.

The cycle of life here was relatively intact, with down and dying trees supplying ample habitat for woodpeckers and a zoo of other snag and rotting-log dwellers, and native soils hosting species of well-drained habitats (not compacted post-cattle clay) such as white and black oaks, mapleleaf viburnum, and rue anemone.

Land Use History and Habitat Quality

Land use history is a primary predictor of habitat quality. The time since abandonment from agriculture is positively correlated with native species diversity and the presence of specialist plant and wildlife species. If your site faced intense disturbance or land use conversion during or after the 1950s, invasive plant species and overabundant deer have likely had an influence on its composition.

Landscapes with a history of recent human disturbance, invasive species colonization, and deer overbrowsing offer both a challenge and an opportunity. It is here that our attention as stewards might be most rewarded, as we guide habitat composition through removals, plantings, and even intentional disturbances of our own.

Invasive species and degraded habitats can seem ubiquitous, random, and overwhelming. By viewing them through the lens of land use history, we can see beyond symptoms to root causes, and consider the effectiveness of various interventions. Should we clear out some multiflora rose beneath the young red maples and farm ginseng in the compacted, post-agricultural clays? Probably not. Would a clearing in these same young woods support wild hazelnut, blackhaw, highbush blueberry, and spicebush? Almost certainly.

Post-Agricultural Soils

Several hundred years ago Europeans arrived in various parts of our region and commenced altering the North American landscape into a more European one. You can hear echoes of this desire to transform the alien into the familiar even in post-colonial place names: New Jersey, New York, New England ... not to mention countless Bristols, Belfasts, Burlingtons, and so on.

In a famous series of dioramas depicting land use change in Harvard Forest in Massachusetts, the viewer is looking out from a rocky knoll. The first diorama shows the pre-colonial forest, clad in tall trees. The second shows a rough clearing in the woods, with a simple cabin and a few livestock. The final diorama depicts a vista of rolling hills denuded of trees, complete with houses and barns, looking very much like a contemporaneous scene from pastoral Europe.

The resemblance to European landscapes is more than superficial. North American soils and plant communities were intentionally altered to better support Old World domesticated plants and animals. Native forest vegetation was cleared, and annual crop fields were plowed, limed, and manured. "Improved" pastures were planted to European grass species like orchard grass and timothy. Even rough pastures, unplowed and often retaining at least some native species, were likely altered by the manure, gut biology, grazing style, and utilization intensity of Old World domesticates like sheep, cows, and goats. Even when New World cultigens such as corn were planted in farm fields, the details of agricultural practice followed European techniques instead of native techniques that favored polycultures and working the soil by hand.

After a few centuries, give or take, of converting the native soils of the Northeast to European-style soils, it is no wonder

From pre-colonial forest to agricultural vista: dioramas depicting the Harvard Forest.

that post-agricultural landscapes are often dominated by Eurasian invasive species. Plants maintain complex, mutualistic relationships with mycorrhizal fungi as well as bacteria and other soil organisms, and when native soils are converted to European-style soils, Old World agricultural weeds prevail.

As these non-native dominated fields undergo succession to post-agricultural

forest, they can maintain soil plow layers, lack native soil organisms, and feature a strong non-native plant presence as well.

The places where we find remnant native plant communities are often in areas too rocky, steep, wet, dry, cold, hot, or otherwise forbidding to have been worth converting to a crop field or an improved pasture. And when we find native plant communities, we are likely in the presence of less-altered native soils as well.

The Story of a Place

A few years ago I was hired to do a botanical survey at a 2,000-acre park in northern New Jersey. Within the park were the flanks and summit of Turkey Mountain, a steep-sided hill of resistant gneiss that had been scoured during the last glaciation and had thin soils over numerous shelves of exposed bedrock.

Along its steep south-facing flank were open glades with groupings of low-flowering herbs interspersed with lichen and moss-clad rock, under widely spaced small trees. Many of the trees were stunted and eccentric. An old cedar clung to one broken rock face, another flat exposure of rock was covered in a drift of prickly pear cactus.

From the earliest spring survey dates to the very end of the season, this flank of Turkey Mountain was our favorite destination. Its open glades and thin woods had all the spring-blooming profusion of a mature forest, but also the mid-summer flowering of a meadow in full glory.

In the early spring, drifts of small herbs arose from the matrix of moss and lichens — early saxifrage, columbine, pussytoes, rue anemone, rock harlequin, and roundleaf ragwort formed a patchwork groundcover in whites, reds, pinks, and yellow. By late spring, prickly pear cactus was blooming along with two rarities, fourleaf milkweed and Virginia snakeroot, with native roses and showy annuals like Venus's looking glass. Mid-summer the glades were rampant with the trio of woodland sunflower, upland boneset, and hoary mountain mint, like a prairie on a hill. And in the late summer and early fall, late purple aster and wavyleaf aster bloomed among the showy seedheads of native grasses — little bluestem, bottlebrush grass, and woodland brome.

During survey work I often inventory 100-square-meter plots. These sound large but really are just a circle with a diameter of about 37 feet — about five paces from the center to the perimeter. One of my plots in the glades had a whopping 54 species in it—the highest number per 100 square meters I have ever recorded in any habitat.

Woodland sunflower (*Helianthus divaricatus*) on a glade at Pyramid
Mountain Natural Historic Area. CREDIT: RACHEL MACKOW

Why so diverse, so abundant, so beautiful?
The story of this place begins with the glaciers
20,000 years ago, dragging their immense
weight and a load of loose till, shaving the
caps off these hills and littering them with
shattered rock. Lichen and mosses colonized
these bare rocks after the retreat of the
glaciers, eventually holding enough debris and
moisture to build thin mats of soil. Only when
these mats reached several inches deep could
woody plants colonize. Sometimes the trees
were thrown in high winds, taking the mat of
soil with them and starting the process anew.
The area remains thinly clad in trees now,
thousands of years after the hill was scalped
by the glaciers. Maybe fire was a force too.
Perhaps vegetation on the dry, thin soils of
the south-facing slope carried wildfires or
Indigenous-set fire up to the summit in the
past as well, sculpting the vegetation, awakening dormant seeds, thwarting the dominance
of forest yet one more time.

Other parts of the story could be told: the
particular character of the potassic gneiss

underlying the glades; the nearby exposure of Franklin marble that may have acted as a refuge for uncommon species that also colonized the glades; and an unknown tale of wild grazers, seed dispersers, and Indigenous habitation.

Climb to the summit of Turkey Mountain with me and I'll show you something surprising. We'll cross a low, fallen stone wall ... and everything changes. The glade flora is gone — no crimson columbines or rare milkweeds — and instead a good portion of the hill is clad in non-native shrubs. As we walk along former farm roads, multiflora rose, barberry, burningbush, and garlic mustard predominate — all invasive species. There are far fewer flowers, and the invasive shrubs form an unappealing thicket. A few hoary mountain mint and smooth rock cress are the only reminders of the glade plant community just over the wall.

A 1930s aerial photograph of the area past the stone wall shows the remnants of a large clearing, probably a farmstead on the hill with its attendant agricultural fields. Probably the area was cleared and used for grazing livestock in the 19th century.

The contrast couldn't be starker. The story of the glades in many ways dates back thousands of years to the last glaciation. Across the stone wall, the story was truncated and is really only a hundred or so years old. Complexity, beauty, and diversity all suffered in the transaction.

Some people argue that invasive species are inevitable, that the more competitive species are destined to take over from the lesser, and to sit back and accept the new normal. But the argument lacks context. Specialist species (as opposed to generalists) and high levels of diversity are deeply linked to the story of a place. Find a place with a deep story, one that hasn't been effaced by land use, and you will find abundance, diversity, beauty. Find a place monopolized by invasive species, and you lose something more than just some names on a plant list. The story of the place has been erased. The character of the low herbs, the crooked trees, the prairie-like summer bloom have been replaced with a different story, that of a great homogenous flora comprised of a limited suite of aggressive generalists, decontextualized species from all continents telling the same story over and over and over again across millions of acres of degraded land.

Natural Disturbances

Disturbance is as important in shaping the character of a natural habitat as geology, climate, and water. Disturbance can take the shape of cyclical or episodic windstorms, volcanoes, fire, or flooding. Disturbance can be caused by the grazing of buffalo (or mastodons), the tunneling of moles, or the periodic incursions of a bark beetle.

Or disturbance can be of modern advent and take the shape of the bulldozer, the boom sprayer, or run-off from I-95.

In the case of disturbances rooted in climatic or geological processes or wildlife behavior, these have often been components of systems across long timescales and the resident flora and fauna have become adapted, even dependent on them.

Indigenous human disturbances can shape habitats as well, and there is ample evidence that very complex, diverse systems ranging from the midwestern prairie to coastal grasslands and oak savannas (on both coasts) were generated by Native American land management practices such as burning, harvesting, and tending.[22]

The types of disturbance caused by modern humans favor different plant species altogether — usually weeds or invasive species that thrive in bare, damaged soils, such as those created by earthmoving equipment, agricultural conversion, rapid erosion, and extensive compaction. These types of disturbance are fairly new in North America and often disfavor native species, not to mention unique assemblages such as those found in old growth forests, eastern grasslands, or pine barrens habitats.

It is important to understand which kind of disturbance your site is experiencing, and what kinds of disturbance it might be adapted to.

The New Jersey Pine Barrens are a classic example of a fire-dependent ecosystem.

Pickering's morning glory (*Stylisma pickeringii* var. *pickeringii*) in the New Jersey Pine Barrens.

Containing a number of endemic and globally rare species, the Pine Barrens is characterized by low-nutrient, sandy soils dominated by evergreen species such as pitch pine and Atlantic white cedar. Open pine woodlands support thick understories of heath species like black huckleberry and lowbush blueberry, and the most open areas host locally to globally rare herbaceous species ranging from turkey beard to Pine Barrens gentian to Pickering's morning glory. The dominant pine species have serotinous cones, opening and shedding their seeds in response to fire. In the absence of fire, both the understory and canopy in the Pine Barrens can grow thick with shading deciduous species such as oaks, and both pines and rare herbaceous species diminish.

The ecological restoration practitioner must therefore consider not just missing plant or wildlife species, not just the redressing of damage done to the soil, but also, in some cases, the restoration of appropriate disturbance regimes that many native plant communities depend on.

Sometimes it is human disturbance that is missing. We are the missing wildlife species. A locally relevant example can be found in Robin Kimmerer's beautiful book *Braiding Sweetgrass.* Her book argues against the conceit that all utilization of wild plant species leads to declines of that species.

Sweetgrass is an important ceremonial species for many Indigenous cultures in the eastern United States, harvested for basketry and smudging. In the book, Laurie, a graduate student studying with Kimmerer, sets up an experiment comparing a plot where traditional harvesting methods for sweetgrass were used against a control plot where no harvest is done. To the shock of the traditional academics on Laurie's review committee:

"The sweetgrass that had not been picked or disturbed in any way was choked with dead stems while the harvested plots were thriving." [23]

The experiment demonstrating the benefit of human harvesting on sweetgrass is corroborated by a map of sweetgrass occurrences, documented by Kimmerer and her student Daniela Shebitz. Blue dots on the map represent places where sweetgrass populations had disappeared, red dots marked historical occurrences that are still extant. It turns out the red dots are clustered around Native communities that utilize the plant. Kimmerer poetically summarizes: "The grass gives its fragrant self to us and we receive it with gratitude. In return, through the very act of receiving the gift, the pickers open some space, let the light come in, and with a gentle tug bestir the dormant buds that make new grass." [24]

I suspect that there are many other species that are dwindling because of a lack of human presence. Many wildcrafters assert that even ginseng, now horribly overharvested, once benefitted from the activity of foragers, who dispersed the seeds of plants they picked into appropriate habitats and replanted sections of roots, reproducing the plant clonally.

CHAPTER 7

Reading the Story of the Land

PLANTS TELL THE STORY of the land in meticulous detail. If only we could read half of what they have to say.

They are the ultimate diagnostic tools for assessing the health of soils and water, the impact of land use history, and the presence or absence of natural disturbances.

I remember my first paying gig as an ecological consultant. I was assisting two mentors of mine in doing forest health measurements, in some large forest blocks in northern New Jersey. Mainly I was holding the "dumb" end of a tape measure and counting blocks on a gridded board obscured by vegetation. But I was thrilled to be out there with two great ecologists, exploring forests unknown to me, finding rare plants and listening for red-shouldered hawks as we did our measurements.

After we finished our first field site, at one of the few remaining strongholds for the golden-winged warbler in New Jersey, we proceeded up a rural road through a settlement of Ramapough Lenape homes,

and parked in a sunny opening at the top of a hill. After the extensive forest we had traversed that morning, something here was very different. I was surrounded by Japanese knotweed, mugwort, and phragmites, in a vast, treeless opening. I had a gut sensation that something here was very wrong.

Soon after, the landscape transitioned to a very nice forest, and we were scrambling up rock outcroppings full of woodland sunflower and down slopes clad in blue cohosh. We were all talking about how diverse the forest was, and how the wild plants were thriving in the relative absence of deer browsing.

I asked about the invaded opening where we parked. The Ford Motor Company had dumped toxic waste including lead, antimony, and arsenic from a nearby auto manufacturing plant into the surrounding landscape.[25] Decades later, this ecological dead zone had been colonized by only a few species of non-native plants.

The plants were telling the story loud and clear.

Plants don't just indicate toxic waste dumps, but every inflection of soil composition, drainage, and disturbance. It's a source of perpetual fascination to me how the plant community subtly shifts from parcel to parcel and hill to hill. Often the same dominant tree species are present, but in differing ratios, and the ground vegetation is subject to very fine gradients, each revealing a different truth in the land.

Common species can be used as indicators that a habitat might contain (or be suitable for) less common species. If I see tuliptree, red oak, and spicebush on a slope, I may start looking for black cohosh and bloodroot. If the site is intact enough, I may even look for ginseng, especially if the geology is limestone or diabase. The common woody plant species suggest a rich, moist soil, and a somewhat protected aspect (the direction a slope faces relative to the sun), that also can support the more specialized herb species.

If, on the other hand, the slope is dominated by chestnut oak, with Pennsylvania sedge and striped pipsissewa in the ground layer, I may start looking upslope for sunny openings that host lowbush blueberry or black huckleberry to forage.

Here's an example from the restoration work we did on our farm. Our property is up on the side of a boulder-strewn hill, mostly forested but with about a two-acre opening that includes our house and nursery. It had been a conventional corn and soy field six years ago when we bought the place. I love our place, but there is one thing we lack — running water of any kind. I've often wished for a stream for any number of reasons, not the least of which as a place to introduce plant species that thrive on the moist, deep soils flanking streams.

Looking at the soy field with an eye to its restoration, one area stood out as a possible location for some of the riparian (streamside) species I wished to plant. It was along the lower fencerow, at the base of the sloped former farm field. Just below it on the slope was a steep, short bank, and a road. I imagined that the road was our stream, the roadbank its sloped sides. Some trees in the fencerow provided a bit more shade than elsewhere in the field, preserving a bit more soil moisture. Also, farm soil had eroded down to this low spot over the years. But what really confirmed the picture for me were some of the weedy species. They had "moist disturbed thicket" written all over them. There was virgin's bower, a native clematis that is often found on sunny shores. Jewelweed, a native annual herb that thrives along riverbanks and in wet meadows. Poison ivy, which prefers moist, disturbed sites and may reach its greatest expression in southern swamps where it radiates off bald cypress knees like a giant shrub. And cleavers, a sprawling weed that clings (gently) to one's pantleg after it goes to seed, asking to be dispersed to more of the good, rich, weedy land that humans prefer to inhabit and create. Even the tree species in the fencerow were good, if not unequivocal, indicators (some trees can be very broad in their habitat requirements): white ash, hackberry, white mulberry, and even slippery elm. Pretty good species for rich, moist, disturbed, riverine habitats.

With these indicator species confirming the potential of the habitat, we set about planting and seeding species that we felt would appreciate the moist, rich, disturbed conditions, and even jokingly referred to the area as our "riparian" planting as if

there was a stream there and not just a country road.

We planted many shrubs; the species we selected thrive in the somewhat sunny conditions along the shores of streams, rivers, and ponds, and the edges of open wetlands. The shrubs also serve as good screening from the road so we could do things that the guy driving by in his pickup truck with the Confederate flag doesn't need to know about.

We planted an edible shrub community of pawpaw, American hazelnut, nannyberry, American plum, Appalachian gooseberry, elderberry, chokeberry and more, with wild persimmon trees interspersed for a

Our "riparian" roadside planting.
CREDIT:
RACHEL MACKOW

later successional element. We introduced herbs like hollowstem Joe Pye, purplestem angelica, and cow parsnip from seed. We augmented with groundnut, an herbaceous vine with edible tubers, and a big sward of bee balm, one of our favorite tea herbs and a riparian species itself. We added many wheelbarrow loads of wood chips, and about five years in, those conventional farm field soils began turning around, the shrubs are maturing, and we began getting our first big harvests of hazelnuts, elderberries, and currants.

Reading Your Landscape's History

Understanding the history of the site you are working with is essential to planning a restoration. Common land uses on many sites include a past of crop agriculture, grazing, orcharding, logging, mining, as well as structures, roads, and parking areas. Often your property will contain a mix of these, sometimes in different sections, sometimes overlaid through time.

Below, I present two distinct modes of inquiry for getting to know the history of your landscape. Like putting together clues in a mystery novel, combining these methods is an exciting and revealing way to immerse in the land.

Historical Aerial Photos and Maps

Here in New Jersey, we have two incredible resources when it comes to interpreting a site's prior vegetative cover and land use.

Because similar resources are available to varying degrees in other states, I will use them as examples here.

The first historical resource is aerial photography. The oldest complete set in New Jersey dates to the 1930s. These are photographs of the entire state taken from the air, and at least one new set is available for every subsequent decade. You'll have to do some searching to find what is available for your state or province. In New Jersey, the Department of Environmental Protection makes these aerials available digitally in a variety of formats. In other states and provinces, these may reside with the DCNR, DEP, or local equivalent. As an alternative to public agencies, an excellent online site is Historicaerials.com. The site is a for-profit venture but casual browsing and printing is free.

In New Jersey we also have the good fortune to have hand-drawn maps rendered between 1870 and 1887 by State Topographic Engineer C.C. Vermeule that portray all of the existing forest cover at the time. This is an incredible vehicle for traveling back in time, as much agricultural land was still actively farmed, and the forest blocks depicted on these maps are generally the woodlots and marginal lands that now comprise the oldest forests in the state. Depending on what state or province you're in, you may have access to similar documents, or to works of historical ecology written by academic researchers.

Food and Medicine
Groundnut
(*Apios americana*)

Groundnut (*Apios americana*) flowers.

Groundnut tubers.

Groundnut tubers sizzled.

Groundnut served. The skin is crispy and the flesh inside resembles chickpeas and potato.

Spikenard
(*Aralia racemosa*)

Spikenard (*Aralia racemosa*).

Spikenard shoots.

Preparing rhizome for making honey spikenard syrup.

Cooking spikenard shoots. They have a very meaty, satisfying flavor.

Shagbark hickory
(*Carya ovata*)

Shagbark hickory (*Carya ovata*).

Shelled hickory nuts.

Drying hickory nutmeat on the woodstove.

Hickory nutmeats on acorn waffles with maple syrup (and Irish butter...).

Wood nettle
(*Laportea canadensis*)

Wood nettle (*Laportea canadensis*).

Wood nettle harvest.

Wood nettle cooked with wild leek foliage.

Crispy and deeply nutritive wood nettle greens.

Ostrich fern
(*Matteuccia struthiopteris*)

Ostrich fern (*Matteuccia struthiopteris*).

Unfurling crosiers.

Fiddlehead harvest.

Fiddleheads sauteed with wild leeks.

Giant Solomon's Seal
(*Polygonatum commutatum*)

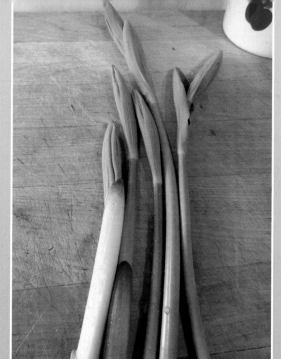

Spring shoots prior to cooking.

Giant Solomon's seal (*Polygonatum commutatum*).

Above: Solomon's seal rhizome.

Right: Making a tincture from Solomon's seal rhizomes.

Oaks
(*Quercus* spp.)

Top left: Glade with chestnut oak (*Quercus montana*) and northern red oak (*Q. rubra*).

Top right: Making acorn flour by removing water-soluble tannins.

Bottom left: Acorn meal and muffin fixings.

Bottom right: Muffins made with chestnut oak flour.

Acorn:
Recipes for the Forgotten Food
25 Nutritious, Delicious & Easy Rec...
to Bring Acorn Back into Your Kit...

Julie M
Eddie

Tall coneflower
(*Rudbeckia laciniata*)

Above: Tall coneflower (*Rudbeckia laciniata*) with bumblebees.

Below: Cooked greens with bacon.

Top right: Basal leaves of tall coneflower in the spring.

Bottom right: Harvesting tender young foliage.

Analyzing the Landscape

It is not just in documents and maps that we can discern the ghosts of prior land use.

Clues to land use history can be found in many places, including the shapes or ages of trees, the presence of stone walls, altered topography, and the presence of persistent old homestead ornamental plants.

A fantastic book on this topic is *Reading the Forested Landscape* by Tom Wessels. It is particularly focused on the New England landscape but it is broadly applicable. I won't attempt to cover all the ground that Wessels does, but here are some especially potent clues I look out for while doing fieldwork and trying to piece together the history of a site.

Indicators of a Remnant

The native herbaceous plants will give you clues about how intact or disturbed a site is. From there, you can proceed to other types of inquiries to try to figure out why.

Indicators of a remnant site with the least disturbance include the presence of high diversity, and niche or specialist native species. There's so much complexity here, so I'm going to winnow down to a few particulars to get you started. We'll focus on forests because most older sites in our area will be forested.

Spring Wildflowers

In high-quality deciduous forests, one can expect good coverage and diversity of

Did these black birches germinate on the same nurse log?

vernal wildflowers such as spring beauty, trout lily, Dutchman's breeches, cutleaf toothwort, bloodroot, foamflower, rue anemone, hepatica, marsh marigold, Canada mayflower, trailing arbutus, wintergreen, partridgeberry, and early sax-ifrage. Seeing three or more of these vernal species in close association, especially in expansive drifts, can be a good sign that the area you are looking at is a remnant natural area to some degree.

Blooming in the spring is a temporal niche well suited to herbs evolved for life under trees. Many of these plants flower and engage in primary photosynthesis during the period when trees haven't yet leafed out. There is less shade, and ample water availability as trees are barely transpiring.

Sometimes you'll see a strip of spring wildflowers along the banks of a stream. Often there are small bands of remnant vegetation along waterways. Just inland may have been pastures or cropfields but the land along the stream was left unmanaged, or trees were deliberately retained for shade for livestock.

Understory Shrubs and Trees

Some woody plants are not canopy trees but thrive in the dappled shade beneath towering giants like tuliptrees, oaks, and maples. These species are rarely the first colonizers of a reverting old field. Instead, they may appear as a forest matures, filling in small gaps beneath the canopy. Therefore, they can be more reliable indicators of forest age than the presence of canopy trees, many of which can act as pioneer species post-agriculture.

An abundance of understory shrubs and subcanopy trees often indicates a forest with a long time since cultivation or perhaps one never converted to open land. Species to look for include mapleleaf viburnum, witch hazel, lowbush blueberry, mountain laurel, hornbeam, hop hornbeam, beaked hazelnut, flowering dogwood, spicebush, striped maple, purple-flowering raspberry, and downy serviceberry.

Sometimes you'll find a linear strip with more understory shrubs and trees, and often more diverse canopy trees, compared to the adjacent forest. This may be a remnant hedgerow that persisted in retaining native soils and plants in a sea of crop or pasture land.

Shade-Tolerant Canopy Trees

Like understory trees, some canopy trees have evolved to be tolerant of the shade of an existing canopy, to grow up beneath it, and to replace it over time. These species include beech, hemlock, and sugar maple. If you find these species dominating the canopy, it is likely that at least a full generation of successional canopy trees has preceded them. Coupled with other clues, shade-intolerant trees in the canopy can indicate a quite old, often less disturbed forest. They might also indicate a lack of fire in locations which may have experienced wildfire or Indigenous fires in the past.

Conservative Plants

Want to go deep? Floristic Quality Assessment is a method initially created to determine the remnant status of a natural area. In short, it ranks all the plants in a state or region's flora on a scale of zero to ten. High-ranking species (C values of 7–10) are habitat specialists that disappear with anthropogenic disturbance that converts or destroys their chosen habitat niches. Medium-ranking species (4–6) are wider in their habitat tolerances as well as their ability to handle medium-intensity disturbance. Low-ranking species (0–3) are generalists that inhabit recently or intensely disturbed sites.

Researchers Bauer, Koziol, and Bever investigated the life history differences between low C value and high C value species. They found that:

> Traits including fast growth rates and greater investment in reproduction were associated with lower C values, and slow growth rates, long-lived leaves and high root:shoot ratios were associated with higher C values. Additionally, plants with high C values and a slow life history were more responsive to mutualisms with mycorrhizal fungi. [26]

Floristic Quality Assessment is covered in detail in the Appendix.

The presence of numerous higher-ranking plant species, with C values from 7 to 10, strongly suggests that your site is a remnant. A remnant is a natural habitat that has not undergone habitat conversion in a very long period. Some remnants, like old growth forests, intact prairies, or undisturbed bogs, may have been in the same habitat type for centuries or even millennia.

Over time you will get to know good indicators for your area and use these as quick clues to how intact or disturbed a site is.

Indicators of Disturbed Habitats

There are many types of anthropogenic disturbance typifying the resource extraction and land use of the last several centuries. Primary among these are cropping, grazing, logging, mining, and residential, commercial or industrial development. Here are some indicators of prior disturbances that can clue you in to historical land uses.

Wolf Trees

You're walking through a forest of straight pole timber when suddenly you come upon a great spreading oak, its boughs radiating low and wide from a girthy trunk. Suddenly the other trees in the forest, while tall and straight, look like lanky teenagers. The old tree is a wolf tree. It originated as an open-grown tree in a pasture, left there by farmers as a shade tree for cattle or other livestock. Frequently, these wolf trees are oaks or hickories, which would have made farmers happy by supplying mast as forage, but other species are possible. In the far north, they may be large, spreading conifers. Wolf trees might also be trees at a property corner or along a property line, retained as markers or found in the no man's land of fencerows. Those spindly teenage trees? They are growing straight because, unlike the branchy wolf tree, they

Oak wolf tree surrounded by younger forest.

established at a similar time to each other, after field abandonment, and are competing for access to the sun.

Logging

Even forests that have never been converted to agriculture have probably been logged at least once, usually numerous times. The frequency, timing, intensity, and methods used in logging may be important factors in how disturbed a forest is now. For example, forests in my area that have been logged in the past two or three decades are very likely to have higher densities of invasive plants in them, simply because these species were so prominent in the surrounding area at the time of disturbance, and because of the spike in deer populations beginning in the middle of the 20th century. Logging remains a very significant factor in our region, with 58% of tree mortality in the northeastern United States directly linked to forestry activities.[27]

Among the clues to logging history can be a network of woods roads, and the presence of old log landings. Sometimes single-species clusters of shade-intolerant species such as tuliptree within forests with a diverse or different canopy can suggest selective logging.

Multi-Trunked Trees

One of the most ubiquitous clues to former logging involves the presence of numerous multi-trunked trees. Nearly all of the forest trees in our flora naturally grow a single trunk, the better to attain the canopy in competitive conditions. When you find forest trees with multiple trunks, it either means they were in an opening or edge since youth, or, as is frequently the case, they have resprouted after being cut. Being top-killed by fire can also have this effect, as can repeated browsing in some species.

After cutting, many trees possess the ability to sprout from the cut stump. These new leaders may pop up in great number, with no apically dominant trunk to suppress this kind of sucker growth. As a result, stump-sprouting trees typically have two or more new trunks.

Oaks are especially good for this kind of sleuthing, as they are valuable timber trees and also ready sprouters. If you're walking through a forest and see several multi-trunked oaks, chances are very high you are witnessing the evidence of prior logging.

Looking at the size of these new trunks can help you guess the time since logging, if you have some experience with guessing tree ages.

Non-Native Vegetation

The presence of old "homestead" plantings in the middle of what seems to be wild forest is a solid indicator of a former house site. Species like periwinkle (*Vinca*) and wisteria are especially common, but many others, including lily-of-the-valley,

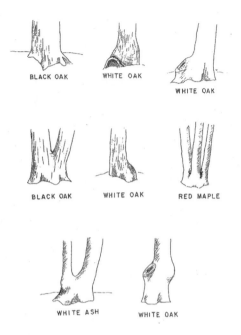

Figure 2. Evidences of stump sprouting. Such trees as these have had their origin as sprouts from stumps left after cutting trees in the past. Sometimes two or more sprouts live as trees. Often all but one die, leaving a deformed base or a rotten hole at the base of the tree that does live.

Evidence of stump sprouting.

Credit: From James Baird's 1956 monograph *The Ecology of the Watchung Reservation, Union County, New Jersey*[28]

bridalwreath spiraea, daylily, and others, can persist for many decades after the domestic setting has been subsumed by forest. Look around and you may find a crumbled foundation with birches clinging to stonemasonry and ferns obscuring the walls.

Also worth looking for are old fruit trees, especially apples and pears, which can be quite persistent. Finding them in significant quantity suggests that your site may have been an orchard in the past.

Mining

I present you with a conundrum. Many of the historically mined and quarried sites, and the land surrounding them, are now among the higher quality forests I know. This is not to suggest that strip mines and other contemporary mineral extraction sites are not heinously disturbed and toxic. Rather, it is the medium-scale iron mine or marble quarry of distant yesteryear that seems to have recovered admirably. Maybe this is because, as destructive as mining is, widespread cropping is even more damaging to soils? Maybe it is because many of these mines were abandoned when no longer profitable, and sold to state and county parks systems early on in the preservation movement? I suspect the early (100+ years ago) release back to nature, at a time when invasive species were highly infrequent, and the deer population was next to zero, may be more to credit than anything about mining itself. On the sites of former mines and quarries, the removal of organic-matter-rich upper soil horizons can create low-nutrient, low-moisture soil conditions which native specialist plants are better adapted to than most invasive non-natives.

There are some incredible online resources showing locations for old mines, as well as some details about the primary resource and time of abandonment. I like using Mindat.org, and also the Mines and Mineral Resources layer in Gaia GPS. That said, many other sources and apps are available.

In the landscape, mines often appear as shallow pits in the ground, especially when they have been filled or are just exploratory pits. Sometimes mineshafts can be found in the sides of hills or descending from pits. Many quarries are filled with water; some ceased operation when groundwater was hit and pumping became prohibitively expensive. If you find a surface mine, try to find a pile of old mine tailings as well. These might have interesting plants growing on them, and if you're even one bit a rockhound, you can collect interesting mineral species in these dumps as well.

Making Your Own Map

When I take on a large botanical survey project, the first thing I do is produce a series of maps. The most important of these feature historical aerials and/or drawings, contemporary aerials, as well as maps for topography and hydrology. I subcontract to a colleague who is a specialist in the use of GIS (Geographic Information Systems) software. However, publicly accessible tools are numerous and serve a variety of purposes. A short list includes:

- Google Earth, easily accessible and with a wide array of features including 3D visualization
- Gaia GPS, which I use in the field and includes geological layers,

mining sites, and private property lines

- Historicaerials.com, for historical aerial imagery from the 1930s to the present
- Macrostrat.org, for maps of rock types
- USDA Web Soil Survey, for soil unit descriptions

In my first pass analyzing a site's map, I'll separate out forested or other wild habitats that were in natural cover in my earliest sources (aerials or drawn maps) from those under agricultural utilization.

These older patches are often the most likely places to find high levels of specialist species, diversity, and structural complexity. Finding these patches will help you to answer a fundamental restoration question: to tend or mend?

Tend or Mend

In looking at all the factors above — soils, water, and land use history — you are also drawing closer to answering a crucial question: is your site a high-quality natural area, or something more disturbed?

Different sites need different restoration approaches. A fundamental choice in restoring any site involves an honest evaluation of the current status of the site. On the one hand, your site may be a natural area remnant with at least some of the components of an intact ecosystem. The

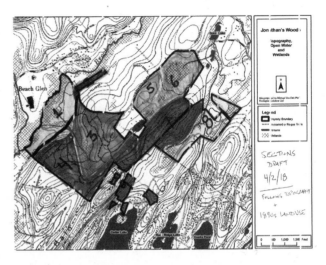

Sketching section boundaries for botanical survey work at Jonathan's Woods County Park.

Credit: Michael Van Clef, PhD, and the Author

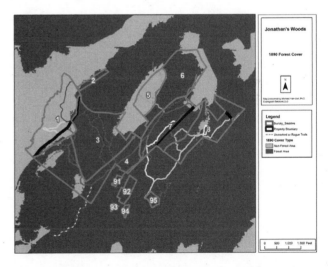

You can see how my survey sections correlate with the forest cover as documented in the 1890s, as well as the topography and hydrology in the map above.

Credit: Michael Van Clef, PhD

restoration process works with these fundamentals rather than seeking to overwrite them in the service of an imposed design. An oak woodland should be tended as such. Other sites are different. They may be so

This botanical survey site was a jewel.

disturbed, or such a blank slate, that they merit reimagining, with recourse to local reference ecosystems offering a palette of possible choices.

I call this fundamental dichotomy "tend or mend."

Some sites are jewels. One year I did fieldwork on a small nature preserve, the signature feature of which is a sheer dolomite cliff over a sinkhole pond. The cliff, monolithic from a distance, is in fact dilapidated into a number of outcroppings, joined by high sloping passes with a deep rich soil. Due to the influence of the limestone geology and the centuries of organic matter accumulation between rocks, boulders, and ledges, the cliff community supports a diversity of plants, including at least a half dozen state-listed rare species. These include a huge population of the evergreen shrub Canada yew, and a decent quantity of the legendary medicinal herb ginseng. Many of the rock faces are completely garbed in moss and lichens, and diminutive ferns like the unusual walking fern creep from one boulder to the next.

Sites like this one need a different approach than an abandoned cornfield or a young woodland dominated by a few successional species. Sites that already host high-quality plant communities deserve tending. These places are elders, survivors that carry essential knowledge from the past. They benefit from our deep interest and involvement, but they don't need us to mastermind them a new future.

Field confirmation of the presence of high- or low-quality communities can be made using a monitoring method called Floristic Quality Assessment. Learn more about it in the Appendix: Assessment and Monitoring Techniques.

High-Quality Sites → Tend

High-quality sites have diverse native-dominated communities including specialist species and intact relationships with aboveground and belowground fauna. Often, these sites are relict sites, spared the worst vagaries of land use and alteration. Soil structure has been relatively undamaged by agriculture or development, and the site's hydrology is functional. Ideally, some level of natural disturbance is still at play, and wildlife aren't terribly out of balance, with appropriate levels of herbivores, abundant pollinators, and maybe even keystone carnivores present.

These sites often speak for themselves, sparking our intuition by communicating beauty and character. To some extent, high-quality sites can also be identified by careful study of historical and contemporary aerial imagery, revealing places that have experienced the least historic disturbance. As I mentioned earlier, when I begin botanical survey projects, I often consult historical aerials as the very first step in planning my approach. I can often discern areas that

merit special attention, or guess which areas are likely to have degraded habitats.

For high-quality sites, plant composition is already fairly set. Canopy dominants, common herbs, and specialist species are all in place. Structure (closed or open, dense shrub layer or none at all) may be malleable but only with careful consideration. While we might augment with a few well-chosen species, the tending of a high-quality site is based around fulfilling the needs of the existing community.

We might ask what degrading factors are present and how can we mitigate them.

- Are there imbalances in the animal community that are negatively impacting the site?
- Are invasive species making inroads, and if so, what are the root causes of their successful spread?
- Are there forms of disturbance *necessary* to the health of the community, and are the right disturbances occurring at the right time — or at all?
- Are there rare or uncommon species that have specialized needs that we can help fulfill?

Harvest and use are by no means off limits in a high-quality community. But they must be attentive to fulfilling the needs of the community and to factors that put the community at risk.

Low-Quality Sites → Mend

Many sites have been degraded by prior land use, by fragmentation, by deer over-abundance and invasive species. Often many degrading factors work in concert. These sites present a huge opportunity for restoration, and have a concomitant need for our intervention.

Often, these sites are rewilding but under marginal conditions. They may lack nearby sources for native seeds, as well as key members of the animal community. They may have plants or animals that are out of balance and may be dominated by a few generalist, weedy, or invasive species. In sites like these, our restoration goal may bear little resemblance to the current condition, as when an abandoned field is converted to a thriving shrubland, or a dense thicket is dramatically thinned to allow the development of a ground layer flora.

These sites need a more extensive approach, mending rather than tending. It is likely that you will want to introduce assemblages of plant species and remove others. Perhaps key disturbances need to be introduced and other, deleterious ones stopped. The full restoration toolkit explicated in this book will be put to work at sites like these. And the restoration target for sites like these will be open to interpretation by you, the restoration practitioner.

In the next section, we'll explore the plant communities that can serve as guides for restoring the land.

Plant Communities and Culturally Useful Plants

A FEW YEARS BACK Rachel and I took a vacation to a region of northeast Pennsylvania dubbed the Endless Mountains. Aerial imagery shows a massive landscape of folded earth with countless ridges running in wave-like ripples, girding boggy valleys.

The country roads are lined with farms and feral fields of meadowsweet, arrowwood, blueberries, and spotted Joe Pye weed. Chokecherries hang heavy with clusters of deep red fruit along the roadsides. So laden with fruit are these bushes that we joked that the Endless Mountains might sink below sea level.

With ample time on our hands, we experimented with new foraged species and dishes. We gathered wild fruits and vegetables, preparing at least one wild food dish every day: wood nettles skillet-fried and formed into a rich green cake; sumac "lemonade"; snacks of wild blueberries; yellow birch-hemlock tea; sheep sorrel salad; and a sauce with two of the largest wild leek bulbs we'd ever seen.

One day we were seeking wild populations of bee balm, which is quite rare in my home state of New Jersey. Deciding to visit a local creek, we walked through a beautiful wooded floodplain towards the water. The herbs were classic forested floodplain species: drifts of wild leek with golden Alexanders, blue cohosh, wood nettle, and trilliums.

When the shade parted I stepped out onto the sunny streamside cobbles where the land was constantly reshaped by water.

Here, bee balm grew in great abundance, with spotted Joe Pye weed, virgins bower, boneset, sensitive fern, stinging nettle, and cardinal flower.

Spotted Joe Pye weed along a Pennsylvania creek.

In the streambed, tussock sedges lay flattened in a downstream direction, and on the opposite bank the dark purple shade of a hemlock forest brought out the colors of the wildflowers along the bank in vivid relief.

Edible and medicinal plants were numerous. I took notes on one stretch of stream a few hundred feet long and tallied wild leek, common milkweed, yellow birch, blue cohosh, boneset, beech, Virginia waterleaf, wood nettle, ostrich fern,

wild mint, bee balm, cutleaf coneflower, common elderberry, basswood, eastern hemlock, and stinging nettle among the species I would gladly use for food or have used as healing herbs.

By contrast, many of the streams and rivers in New Jersey where I live host degraded plant communities. They've borne the brunt of colonial agriculture, milldams, logging, and industrial age exploitation. Excessive deer browse and colonization by invasive plant species has further diminished diversity. Despite this they retain in large part their rich soils and generous moisture. The wild food palette that can grow in such places is prodigious. If I was restoring one of these streams, I'd consider plants from the Endless Mountains stream listed above, and I could add some fantastic staple starches such as groundnut and Jerusalem artichoke, delicious fruits such as persimmon, pawpaw, and American plum, and sugar in the form of the fast-growing, disturbance tolerant silver maple and box elder trees.

Returning home to the familiar degraded streams and rivers in New Jersey, it was clear that they have so much potential for restoration. In this section, we'll explore habitats and their associated plant communities. Like the Endless Mountains stream described above, the reference communities for each habitat reveal a rich palette of edible, medicinal, and cultural species that thrive in each.

CHAPTER 8

Plant Communities

PLANT COMMUNITIES are living aggregates that recur in ever-varying patterns across the land.

What determines the composition of plant communities? The complex interplay of many factors: climate, microclimate, geology, hydrology, aspect, slope, disturbances, land use history, biological legacies, and more. Because these factors are revealed in silent glory by the diversity of plants in a habitat, habitats that lack plant diversity are missing a fundamental part of their stories. They can suffer from a sameness similar to that of a strip mall or a suburban development. Part of ecological repair may be allowing a place to express its story again, after a history of plowing, over-grazing, erosion, or other factors that literally erase the character of a place.

Why does a plant belong to a community? Each species has requirements and tolerances: for soil moisture, necessary sunlight, the availability of certain nutrients. Plants with similar needs and tolerances tend to grow in similar places. Over time, they have become interknit in profound ways. Non-plant members of the community emerge and play vital roles. These include soil fungi that allow plants to thrive, and pollinators and birds that aid in reproduction and seed dispersal. We see clusters of similar species co-occurring over and over in space, though these communities are ever-changing and not firmly bounded.

A plant community can *perpetuate* the conditions that allow for its existence by modifying soils, altering temperatures at the ground level, and modulating sun exposure.

Most sites contain more than one plant community type, as a riparian corridor grades to a mesic slope, or an upland forest opens to a ridgeline glade. Even if you're working within a small site, chances are you'll have small micro-habitats — the area under a shade tree in an otherwise sunny yard, for instance, or the steep dry bank of

An upland oak and heath forest community.

those close to home, and can be extrapolated, in degrees, to apply to a wide swath of eastern North America, including parts of Ontario, Quebec, and the Canadian Maritime provinces, and in varying degrees as far west as the Great Plains.

There are many different complex ways to classify the enormous diversity of plant communities. I think it's impossible to do so in a satisfactory way given all the variables involved, and many existing classification systems break down when I try to apply them to real places.

I've tried a somewhat different approach in describing the plant communities below. I pose a few basic questions. These are questions I ask when doing fieldwork or visiting a new place, and I believe you can do so as well. How wet is the site? How flat or steep? Sunny or shaded? Rocky or not? Is the bedrock nutrient poor (e.g. granite) or something richer (e.g. limestone)? How close to the coast am I, and how far north or south?

In addition to asking these questions, I look for indicator species as I do fieldwork. These can be common plants (even weeds), or very specialized ones, which help answer basic questions about soil moisture levels, sun exposure, geology, and the like. With every plant community described below, I've included a list of indicator species.

My system is necessarily flawed. The incredible diversity of plant assemblages

a roadside fringing an open marsh. Allow the communities below to evoke places of any scale, and be open to the diversity that comes from the complex interdigitation of wet and dry, sun and shade, sand and clay.

In describing plant communities for the Northeast, I'm wary of the perils of over-generalization. I've chosen to base my descriptions on the actual habitat types I'm most familiar with, as a field botanist working primarily in New Jersey, and as an active observer throughout the Northeast, Mid-Atlantic, and beyond. Rather than conflate the tremendous variation in our biome, my habitat palettes are based on is impossible to adequately represent in one classification scheme. But my hope is that it's easier to use and understand than the scientific community's classification systems out there. Therein lies its utility as a rough starting point, a foundation if you will, for appreciating the complexity and potential of your own land, the site you are working with.

The plant communities described below will help you to determine what other plant species your site can support, whether you plan to introduce species by planting, seeds, or even by restoring appropriate conditions and hoping that they will recruit on their own. Each plant community description contains a list of edible, medicinal, and other cultural species that tend to be found in that community. The following chapter, Chapter 9: Plant Species, describes each of these species in detail.

Each plant mentioned in the community descriptions has its own native range. Some of these ranges are vast. Red maple, for example, grows native from Quebec to southern Florida, and west to Minnesota and eastern Texas. Others are more geographically limited. Beach plum, for example, grows only in the United States, ranging from Maine to Virginia in coastal states only.

I encourage you to utilize the information below in the following manner. Read the community descriptions below, find some attributes or indicator species in common with your land, and then consider the other plants in the community as possible choices to repair or diversify your site with. If you find a plant of interest, read its description in Chapter 9: Plant Species to get a sense of its cultural uses and basic form. Then, check range maps such as those published online by the Biota of North America Project (web search "Bonap" and the plant name), or the USDA PLANTS database (found at https://plants.usda.gov). Both of these resources cover Canada as well, which does not have a complete native plant range map resource of its own.

Why bother considering plant communities? Why not skip ahead to the next section and pick out all the cool plants? It's a matter of scale. If what you're looking for is a garden, where you will attend to the needs of each of your plants by creating suitable conditions (rock garden, or deep pile of compost), weeding, watering, fertilizing, and so on, you can within reason grow just about any native plant you want. But we need to scale up if ecological repair is our goal. To plant across yards, parks, woodlots, farms, and roadsides, we'll need our plantings to persist without the intensive care of gardening. Matching plants to pre-existing conditions, and doing so with groups of species at once, is our best bet for restoring vast swaths of damaged land.

PLANT COMMUNITY: RIPARIAN FORESTS

Walk with me towards the whitened trunks of sycamores, towards the bright horizon where we know the river flows but cannot yet see it because of the high bank carved by its rushing waters. Watch out for the stinging nettles growing in the loamy soil that compresses softly beneath our feet. Nettles and poison ivy grow lustily in these fertile soils, as do the giant silver maples and the ungainly box elders. Every year I visit this place in early April to harvest a few meals of ostrich fern fiddleheads before they unfurl into giant fronds. In additions to big colonies of these antediluvian ferns, there are some other surprises here, including a few rare plant species which are common to the north and west, and are found along this river corridor but have not been successful inland. There's also a lot of Japanese knotweed and mugwort, not a few old tires, and some dimension lumber that washed down in the last flood from someone's incomplete home improvement project. Welcome to the riparian forest.

Marked by the easy accessibility of water, and featuring gradations of species with varying tolerance for inundation, riparian forests can be very diverse systems in terms of plant species and structure. Soils along waterways are frequently rich and deep, built partially from the transportation of organic materials from upland habitats delivered during flooding.

Plants of river and stream banks grow here — species that tolerate occasional flooding but otherwise enjoy fertile, relatively well-drained loamy soil. For *wetland* communities comprised of plants that thrive in regularly inundated conditions, see the *Forested Wetlands* section, below.

The native plant communities of river and stream banks are exemplary in their food value. Riparian forests are the bread basket of wild habitats. Soils are fertile, light can be plentiful due to the treeless zone (at the very least, the river itself is treeless), and species that require ample moisture can occupy hydrological zones where water availability is most optimal for them.

Riparian plants are also tough: they persist in an environment characterized by inundation, violent flooding, and subsequent soil disturbance. For this reason, riverbanks can be extremely diverse native habitats, but are often also the first places to be overrun by invasive species and other new arrivals to the flora. In fact, it is posited that in the deep past, the earliest angiosperms (flowering plants) may have occupied and thrived in riverine systems, while more stable habitats inland remained dominated by gymnosperms.

Historically, Indigenous settlements in our region were sited near waterways, and this pattern was mirrored in early European settlement patterns. The rich soils, diverse flora, abundant fauna, and ease of transportation make these choice spots for human habitation. Much of the choicest land near rivers has long been converted to agriculture, used for mills, or developed industrially. It is difficult to find large areas of intact floodplain forest to use as reference communities for restoration.

Indicator Species: Woody Plants. Silver maple, hackberry, basswood, box elder, black walnut, hemlock, river birch, sycamore, beech, yellow birch, spicebush, elderberry.

Indicator Species: Herbaceous Plants
Cleavers, jewelweed, wood nettle, Japanese knotweed,* mugwort,* chickweed,* stinging nettle,* lesser celandine.*

**Indicates non-native species.*

Table 8.1: Cultural Species of Riparian Forests

Common	Latin	Form	Primary Use
Red Maple	*Acer rubrum*	Tree	Syrup
Silver Maple	*Acer saccharinum*	Tree	Syrup
Wild Onion	*Allium canadense*	Herb	Edible bulb and tops
Wild Leek	*Allium tricoccum*	Herb	Edible bulb and foliage
Purplestem Angelica	*Angelica atropurpurea*	Herb	Medicinal root
Smooth Serviceberry	*Amelanchier laevis*	Small Tree	Edible fruit
Hog Peanut	*Amphicarpaea bracteata*	Herb	Edible legume
Groundnut	*Apios americana*	Vine	Edible tuber
River Cane	*Arundinaria gigantea*	Grass	Craft/utilitarian materials
Wild Ginger	*Asarum canadense*	Herb	Root tea
Pawpaw	*Asimina triloba*	Small Tree	Edible fruit
Shellbark Hickory	*Carya laciniosa*	Tree	Edible nuts
Hackberry	*Celtis occidentalis*	Tree	Edible fruit
Hawthorn	*Crataegus* spp.	Small Tree	Medicinal flowers and fruits
Leatherwood	*Dirca palustris*	Shrub	Craft/utilitarian materials
American Beech	*Fagus grandifolia*	Tree	Edible nuts
Honey Locust	*Gleditsia triacanthos*	Tree	Sweet pulp
Jerusalem Artichoke	*Helianthus tuberosus*	Herb	Edible tubers

Table 8.1: Cultural Species of Riparian Forests (*Continued*)

Common	Latin	Form	Primary Use
Virginia Waterleaf	*Hydrophyllum virginianum*	Herb	Edible foliage
Butternut	*Juglans cinerea*	Tree	Edible nuts
Black Walnut	*Juglans nigra*	Tree	Edible nuts
Wood Nettle	*Laportea canadensis*	Herb	Edible foliage
Spicebush	*Lindera benzoin*	Shrub	Spice and tea plant
Starry Solomon's Plume	*Maianthemum stellatum*	Herb	Edible shoots
Ostrich Fern	*Matteuccia struthiopteris*	Fern	Edible shoots
Bee Balm	*Monarda didyma*	Herb	Tea from foliage and flowers
Red Mulberry	*Morus rubra*	Tree	Edible fruit
Giant Solomon's Seal	*Polygonatum commutatum*	Herb	Edible shoots
American Plum	*Prunus americana*	Shrub	Edible fruit
Swamp White Oak	*Quercus bicolor*	Tree	Edible nuts
Tall Coneflower	*Rudbeckia laciniata*	Herb	Edible greens
Black Elderberry	*Sambucus canadensis*	Shrub	Edible and medicinal fruit and flowers
Basswood	*Tilia americana*	Tree	Cordage from bark
Slippery Elm	*Ulmus rubra*	Tree	Medicinal bark
Nannyberry	*Viburnum lentago*	Small Tree	Edible fruit
Blackhaw	*Viburnum prunifolium*	Small Tree	Edible fruit

PLANT COMMUNITY: RICH MESIC FOREST

Moving uphill from the stream corridor in the valley, we find ourselves in the protected cove of a bouldery north-facing slope, shielded from the direct intensity of the sun by the crest of the hill way above. Look around and you'll see understory plants with big broad leaves, generous solar panels for absorbing the sun's dappled light. The soils are deep and cool, with humus filling the space between slow-weathering boulders. This is one of the places where forest herbs thrive, sometimes multiple tiers of them beneath a complex woody understory and canopy. The slope gets really steep here for a moment, but I want to reach that flat up above with the swards of maidenhair fern and wild ginger. There were some trilliums down below and I want to see if there are more here.

This is the mesic forest. Mesic, meaning moderately moist — not too wet, not too dry. Soil conditions in mesic forests are moist but well-drained. Plentiful organic matter holds moisture and nutrients, while good soil structure, and factors like stoniness and slope, guarantee adequate drainage. Mesic forests contain conditions amenable to many species, and have some overlap with riverine systems in terms of species composition.

Mesic forests over nutrient-rich geology are termed rich mesic forests. Some geological substrates that support rich mesic forests include limestone, diabase, basalt, amphibolite, and calcareous shales. Related technical names include basic mesic forest in the Piedmont, and cove forest in the Appalachian Mountains.[29]

If riparian areas are the bread basket, mesic forests, especially rich mesic forests, contain a medicine chest of plant species. They contain many powerful medicine plants, including ginseng, goldenseal, black cohosh and others. These species are under threat due to wild collecting for international and domestic markets. Therefore, there is both a conservation incentive and an economic incentive to grow them, especially in naturalistic (wild-simulated) or restored conditions that produce authentic wild-type plants. Plants grown in natural

conditions may be more medicinally efficacious, can fetch a higher market price, and, in the case of conservative species, may be the only or best way to produce these species sustainably.

In my area, indicators of rich mesic forest include tuliptree, shagbark hickory, northern red oak, beech, white ash, and basswood in the canopy; spicebush, blackhaw, and mapleleaf viburnum in the understory; and a diverse herbaceous community in the ground layer. Herbs like maidenhair fern, wild geranium, roundlobe hepatica, wild ginger, black cohosh, bloodroot, purple node Joe Pye weed, and perfoliate bellwort are good indicators for this habitat. There is significant overlap between mesic forest species and those of alluvial terraces, which are also rich, moist, well-drained habitats. Look for species like wild yam root, tuliptree, and spicebush in both places.

Indicator Species: Woody Plants. Tuliptree, shagbark hickory, basswood, sugar maple, black birch, tupelo, hop hornbeam, northern red oak, white ash, white oak, flowering dogwood, hornbeam, bladdernut, winged euonymus,* multiflora rose,* Japanese barberry.*

Indicator Species: Herbaceous Plants.
Virginia stickseed, white avens, white wood aster, blue heartleaved aster, wreath goldenrod, mayapple, cleavers, wrinkleleaf

goldenrod, Christmas fern, common
speedwell,* garlic mustard,* chickweed,*
Japanese honeysuckle.*

Indicates non-native species.

Table 8.2: Cultural Species of Rich Mesic Forests

Common	Latin	Form	Primary Use
Red Maple	*Acer rubrum*	Tree	Syrup
Sugar Maple	*Acer saccharum*	Tree	Syrup
Black Cohosh	*Actaea racemosa*	Herb	Medicinal rhizome
Nodding Onion	*Allium cernuum*	Herb	Edible bulb and tops
Wild Leek	*Allium tricoccum*	Herb	Edible bulb and foliage
Smooth Serviceberry	*Amelanchier laevis*	Small Tree	Edible fruit
Hog Peanut	*Amphicarpaea bracteata*	Herb	Edible legume
Wild Sarsaparilla	*Aralia nudicaulis*	Herb	Medicinal root
Spikenard	*Aralia racemosa*	Herb	Edible shoots
Black Birch	*Betula lenta*	Tree	Bark tea
Cutleaf Toothwort	*Cardamine concatenata*	Herb	Edible foliage
Shagbark Hickory	*Carya ovata*	Tree	Edible nuts
Blue Cohosh	*Caulophyllum thalictroides*	Herb	Medicinal root
Redbud	*Cercis canadensis*	Small Tree	Edible blooms and pods
Spring Beauty	*Claytonia virginica*	Herb	All parts edible
Stoneroot	*Collinsonia canadensis*	Herb	Medicinal root
American Hazelnut	*Corylus americana*	Shrub	Edible nuts
Beaked Hazelnut	*Corylus cornuta*	Shrub	Edible nuts
Hawthorn	*Crataegus* spp.	Small Tree	Medicinal flowers and fruits
Dittany	*Cunila origanoides*	Herb	Culinary herb
Wild Yam Root	*Dioscorea villosa*	Vine	Medicinal rhizome
American Persimmon	*Diospyros virginiana*	Tree	Edible fruit
Purple Node Joe Pye	*Eutrochium purpureum*	Herb	Medicinal root
American Beech	*Fagus grandifolia*	Tree	Edible nuts

Table 8.2: Cultural Species of Rich Mesic Forests (*Continued*)

Common	Latin	Form	Primary Use
Wild Strawberry	*Fragaria virginiana*	Herb	Edible fruit
Wild Geranium	*Geranium maculatum*	Herb	Medicinal root
Witch Hazel	*Hamamelis virginiana*	Shrub	Topical medicinal
Goldenseal	*Hydrastis canadensis*	Herb	Medicinal rhizome
Virginia Waterleaf	*Hydrophyllum virginianum*	Herb	Edible foliage
Spicebush	*Lindera benzoin*	Shrub	Spice and tea plant
Tuliptree	*Liriodendron tulipifera*	Tree	Craft/utilitarian materials
Tupelo	*Nyssa sylvatica*	Tree	Edible fruit
Sweet Cicely	*Osmorhiza claytoni*	Herb	Tea from root
Aniseroot	*Osmorhiza longistylis*	Herb	Tea from root and foliage
Violet Wood Sorrel	*Oxalis violaceae*	Herb	Edible green
Ginseng	*Panax quinquefolius*	Herb	Medicinal root
Mayapple	*Podophyllum peltatum*	Herb	Edible fruit
Solomon's Seal	*Polygonatum pubescens*	Herb	Edible shoots
American Heal-All	*Prunella vulgaris* var. *lanceolata*	Herb	Medicinal tops
Black Cherry	*Prunus serotina*	Tree	Medicinal bark
Chokecherry	*Prunus virginiana*	Shrub	Edible fruit
White Oak	*Quercus alba*	Tree	Edible nuts
Northern Red Oak	*Quercus rubra*	Tree	Edible nuts
Prickly Gooseberry	*Ribes cynosbati*	Shrub	Edible fruit
Missouri Gooseberry	*Ribes missouriense*	Shrub	Edible fruit
Appalachian Gooseberry	*Ribes rotundifolium*	Shrub	Edible fruit
Black Locust	*Robinia pseudoacacia*	Tree	Edible flowers
Carrion Flower	*Smilax herbacea*	Vine	Edible shoots
Basswood	*Tilia americana*	Tree	Cordage from bark
Slippery Elm	*Ulmus rubra*	Tree	Medicinal bark
Blackhaw	*Viburnum prunifolium*	Small Tree	Edible fruit

PLANT COMMUNITY: UPLAND OAK FOREST

We're walking up the rocky slope, steep enough that our ankle muscles strain as we take big obtuse paces from boulder to boulder. The chestnut oaks, with their deeply ridged gray bark, look pretty lithic themselves, as if made of a hybrid of rock and tree. Herbaceous plants have gotten a bit more scant here, and the oak leaf litter is deep and persistent. Among the turfs of Pennsylvania sedge we find diminutive heaths such as striped pipsissewa, shinleaf, and wintergreen. When we start clambering through the mountain laurel thicket ahead, please watch that you don't get poked in the eye. Mountain laurels have a way of holding onto old dead stems that seem to meander and jut out ... at eye level.

Often found on steep, rocky slopes, upland oak forests have excessively drained soils. Those derived from felsic (silica-rich) rocks such as granite and some gneisses have poor soils — low in essential plant minerals and acidic in pH. On stony slopes, only a thin layer of humus lies over glacial till or exposed bedrock. Tree and shrub roots grasp boulders like woody claws, following the rounded contours of stone and seeking cool moist conditions deep in the earth.

Here we find species adapted to low-nutrient conditions, to high soil acidity, to episodic dryness, and to fire, which not long ago coursed up the side of many hills swallowing thatch, duff, downed wood, and living plants alike.

Very similar communities are found in pinelands and other excessively drained sandy areas found on the coastal plain (see *Sandy Pine and Oak Forests*, below).

In much of our area, gravelly, stony, and sandy areas are dominated by a combination of oaks, pines (sometimes), and a number of heaths. The latter are also known as ericaceous species and include blueberries, huckleberries, wintergreen, and mountain laurel. Primarily shrubs, often short in stature, they dominate the understory beneath the oaks. The lowbush blueberries and huckleberries spread by stolons, forming colonies at shin and knee height. They transmit fire so well they are sometimes called, as a group, "kerosene bush."[30] They benefit from the clearing effects of fire on understory competitors as well as canopy trees, and are capable of prolific resprouting from underground stems. Practically speaking, this means that acidic uplands are often gloriously bountiful in berry plants, and management with fire is used in some places to make the harvest even more abundant. Combined with the sheer volume of food produced by the oaks in the form of acorns, this makes upland oak forests a potentially excellent source of fruits and starches.

Somewhat moister or more nutrient-rich upland oak forests will grade towards mesic forests in composition. Practically speaking, this usually means that shrubs will diversify from just heaths and include species such as witch hazel or mapleleaf viburnum. The herb layer may become more diverse, with species such as white wood aster, wreath goldenrod, and wild sarsaparilla more common. Among the oaks, dominance may shift from chestnut oak towards other oaks such as northern red oak and white oak.

In the absence of fire, non-fire-tolerant species such as beech, red maple, and black birch often form a subcanopy in contemporary upland oak forests. In the long term, these tree species may cast shade that limits the ability of oak to reproduce, and may suppress the characteristic understory flora of upland oak forests. These non-fire-tolerant trees are responding to a change in the disturbance regime that once helped shape oaken forest communities. Some land stewards may choose to reintroduce fire in these places to manage in favor of oaks and heaths.

Indicator Species: Woody Plants. Chestnut oak, pignut hickory, red cedar, black birch, Virginia pine, downy juneberry, chokecherry, black huckleberry, lowbush blueberry, mountain laurel, mapleleaf viburnum, red maple.

Indicator Species: Herbaceous Plants. Silverrod, Pennsylvania sedge, Canada mayflower, hay-scented fern, common blue violet, whorled loosestrife, downy Solomon's seal, striped pipsissewa, Indian cucumber root, wild sarsaparilla, sessile-leaved bellwort, cowwheat, wreath goldenrod, partridgeberry, American pennyroyal, sheep sorrel,* butter-and-eggs,* oxeye daisy.*

* *Indicates non-native species.*

Table 8.3: Cultural Species of Upland Oak Forests

Common	Latin	Form	Primary Use
Red Maple	*Acer rubrum*	Tree	Syrup
Downy Serviceberry	*Amelanchier arborea*	Small Tree	Edible fruit
Wild Sarsaparilla	*Aralia nudicaulis*	Herb	Medicinal root
Black Birch	*Betula lenta*	Tree	Bark tea
American Chestnut	*Castanea dentata*	Tree	Edible nuts
Sweetfern	*Comptonia peregrina*	Shrub	Tea
Hawthorn	*Crataegus* spp.	Small Tree	Medicinal flowers and fruits
Wild Strawberry	*Fragaria virginiana*	Herb	Edible fruit
Wintergreen	*Gaultheria procumbens*	Subshrub	Edible fruit and foliage

Table 8.3: Cultural Species of Upland Oak Forests (*Continued*)

Common	Latin	Form	Primary Use
Black Huckleberry	*Gaylussacia baccata*	Shrub	Edible fruit
Witch Hazel	*Hamamelis virginiana*	Shrub	Topical medicinal
American Pennyroyal	*Hedeoma pulegioides*	Herb	Tea from foliage
Indian Cucumber Root	*Medeola virginiana*	Herb	Edible rhizome
Tupelo	*Nyssa sylvatica*	Tree	Edible fruit
White Pine	*Pinus strobus*	Tree	Medicinal foliage, resin, and pollen
Virginia Pine	*Pinus virginiana*	Tree	Medicinal foliage, resin, and pollen
Smooth Solomon's Seal	*Polygonatum biflorum*	Herb	Edible shoots
Black Cherry	*Prunus serotina*	Tree	Medicinal bark
White Oak	*Quercus alba*	Tree	Edible nuts
Chestnut Oak	*Quercus montana*	Tree	Edible nuts
Northern Red Oak	*Quercus rubra*	Tree	Edible nuts
Sassafras	*Sassafras albidum*	Tree	Root for beverage
Lowbush Blueberry	*Vaccinium pallidum*	Shrub	Edible fruit

PLANT COMMUNITY: GLADES

When we climb to the top of the ridge, we see that sheets of bedrock are exposed, and plants grow in swards between stone, like little gardens. Trees are found rooted in fissures in the rock. Thin soil, bedrock and gravels alike are clad in a crust of lichens and moss. Soft domes of white cushion-moss alternate with the coral-like reindeer lichen that crunches underfoot. Trees become sparser towards the summit, and are eccentrically shaped due to exposure to wind and ice. Much sun reaches the ground layer plants.

These habitats are called glades.

This one is on fairly nutrient-poor rock. Along with the stunted chestnut oaks, scraggly pitch pines lean into the view, which stretches across a panorama of forested hills. Pink lady's slipper orchids are nestled in the duff near some scrub oaks. The whole summit is ringed with black huckleberry, with its deep red bell-shaped flowers set off against dark grey stems.

Little bluestem is bending in the wind, and if we came back up here in winter we'd see how wind and ice prune the trees and give them their unusual, bonsai-like shapes.

Glades are long-term, relatively treeless openings within otherwise forested habitat. These days, they are most often the products of the harsh conditions brought about by thin soils over bedrock outcroppings. For example, plants of granite outcrops in the Appalachians experience extreme temperatures. According to Timothy Spira, "[h]igh solar radiation and the absorption of heat by the rock surface result in summer temperatures that commonly exceed 120°F."[31]

In the past, glades might have been maintained by wildfire, by Indigenous burning, and by the now-extinct Pleistocene megafauna that once roamed our area. Uplands in general may have been more glade-like in the past, but glades are now found in areas where extensive exposed bedrock preempts development of a closed canopy forest.

Glades at their best borrow from both upland forest and grassland floras, adding their own distinctive species. Flowering can last throughout the growing season due to this diversity. Glades are a fantastic habitat for small fruit trees (hawthorns, serviceberries, chokecherries, blackhaw), and fruiting shrubs (Aronia, black huckleberry, lowbush blueberry), which tend to produce abundantly given the paucity of canopy tree cover. The trees that are present — chestnut oak, red oak, pignut hickory, red cedar, black birch, black cherry — are often excellent food trees in their own right, and more spreading and orchard-like in structure than in a closed canopy forest.

In some ways, glades are not even that different from fruit orchards, except that instead of shorn grass beneath the widely spaced and spreading trees, they have a diverse ground layer of shrubs and herbs.

One difference worth noting between natural glades and possible imitations, where we might try to recreate such habitats, is that the extremes of the habitat in real glades (dryness, wind, ice damage) limit many otherwise aggressive species from taking hold. Instead, small herbs and grasses, and specialized shrubs and sub-shrubs, can persist for long periods, because they are adapted to tolerate such adverse conditions. Put these same plants in a posh planting bed and some glade species will thrive, whereas others may need constant rescue from overtopping by weeds and by other perennials. Worth keeping in mind before planting some prickly pear cactus, New Jersey tea, or columbine next to that ostrich fern in your food forest!

Indicator Species: Woody Plants. Chestnut oak, pignut hickory, red cedar, black birch, Virginia pine, downy juneberry, chokecherry, black huckleberry, lowbush blueberry, mountain laurel.

Indicator Species: Herbaceous Plants. Sheep sorrel,* little bluestem, poverty oat grass, tufted hairgrass, silverrod, silvery sedge, wild sarsaparilla, cowwheat, Venus' looking glass, rock harlequin, American pennyroyal, wild columbine, pussytoes.

* *Indicates non-native species.*

Table 8.4: Cultural Species of Glades

Common	Latin	Form	Primary Use
Nodding Onion	*Allium cernuum*	Herb	Edible bulb and tops
Downy Serviceberry	*Amelanchier arborea*	Small Tree	Edible fruit
Running Serviceberry	*Amelanchier stolonifera*	Shrub	Edible fruit
Wild Sarsaparilla	*Aralia nudicaulis*	Herb	Medicinal root
Black Chokeberry	*Aronia melanocarpa*	Shrub	Edible fruit
Black Birch	*Betula lenta*	Tree	Bark tea
New Jersey Tea	*Ceanothus americanus*	Shrub	Medicinal root
Hackberry	*Celtis occidentalis*	Tree	Edible fruit
Hawthorn	*Crataegus* spp.	Small Tree	Medicinal flowers and fruits
Dittany	*Cunila origanoides*	Herb	Culinary herb
Wild Strawberry	*Fragaria virginiana*	Herb	Edible fruit
American Pennyroyal	*Hedeoma pulegioides*	Herb	Tea from foliage
Eastern Prickly Pear	*Opuntia humifusa, O. cespitosa*	Cactus	Edible fruit and pads
Virginia Pine, Pitch Pine	*Pinus virginiana, P. rigida*	Tree	Medicinal foliage, resin, and pollen
Appalachian Sand Cherry	*Prunus susquehanae*	Shrub	Edible fruit
Black Cherry	*Prunus serotina*	Tree	Medicinal bark
Chokecherry	*Prunus virginiana*	Shrub	Edible fruit
Hoary Mountain Mint	*Pycnanthemum incanum*	Herb	Tea from foliage
Narrowleaf Mountain Mint	*Pycnanthemum tenuifolium*	Herb	Tea from foliage
Chestnut Oak	*Quercus montana*	Tree	Edible nuts
Lowbush Blueberry	*Vaccinium pallidum*	Shrub	Edible fruit
Blackhaw	*Viburnum prunifolium*	Small Tree	Edible fruit

PLANT COMMUNITY: MEADOWS AND GRASSLANDS

We've been travelling together a lot, these past few pages. Let's stop by my house and take a break over a cup of tea.

Well, since you're here, why don't we take a look at the meadow. You're not tired of plants yet, right?

When we moved here eight years ago, what is now a meadow was a farm field that grew alternating crops of corn and soybeans. The first year we let it go, the young meadow grew a combination of naturalized annuals, agricultural weeds, and a few precocious native perennials. There was a lot of horseweed and Indian tobacco! Blue heartleaf aster blew in from the adjoining woods edge, and Virginia stickseed, my least favorite native plant, was carried in by animals. Two seasons later, goldenrods and wild bergamot were creating showy displays on either end of the summer.

Finished your tea? We'll walk up past our plant nursery's propagation area and into the meadow. Of all the habitats on our land, this has been the most fun and interesting to watch. Partially because we didn't plant any of it, and seeded only in a casual way — if I had a handful of seed left over from nursery seedings, I might toss it into the wind in the direction of the meadow. The tall spikes of blazing star, the golden sheafs of Indian grass, the glowing purple New England Asters, the gray seedheads of wild bergamot, the smatterings of mountain mint and pearly everlasting — they chose their own way here and found the spots best suited to them. Eight years in, the former farm field is highly diverse and almost shockingly beautiful. Credit to us is minimal other than installing the deer fence, and keeping up with annual maintenance mowing. And some invasive plant removals.

In eastern North America, meadows and other sunny open habitats (grassland, savanna, and prairie all vary by degree in composition or are terms used regionally) are usually caused by some disturbance, and maintained as open habitats only through the recurrence of disturbance. Thus, many meadows are transient habitats — abandoned farm fields and empty lots that were once cleared of trees and are presumably headed back to forest at some point.

The plants of meadows range from rapidly colonizing weeds and generalists to a flora as long-lived and complexly interrelated as that of the forest itself. In the midwestern tallgrass prairies we can speak of old growth grasslands in the same way that we speak of old growth forests. In fact, the former are rarer, with only an estimated four percent of tallgrass prairie remaining, most of which is in the Flint Hills region of Kansas and Missouri. East of the Mississippi, the percentage is less than one percent.

In addition to the prairies found primarily from Illinois westward, there are grassland communities in eastern North America including ridgetop glades on shallow soils, limestone barrens and oak openings, ice-scoured river cobbles, sandy back dune habitats, areas of pine barrens with frequent fire return intervals, and coastal grasslands once maintained by Indigenous burning practices.

As a generalization, non-forest plants form persistent communities in areas of adverse xeric or waterlogged conditions, especially over unusual geologies that may aid in arresting succession — and may have limited agricultural exploitation of the site in colonial times.

A study of remnant grassland, meadow, and savannah habitats in Pennsylvania concludes that: "[R]emnants have declined from an estimated 230 to 240 square miles around the time of European settlement (0.5% of the state's total land area) to less than 1 square mile today, a 99.6% decline, which continues and is even accelerating at many sites."[32]

So a small differentiation is in order here. On the one hand, our region hosts a limited but very important number of remnant grasslands, often of high quality and hosting regionally and even globally rare species. On the other hand, and much more commonly, we have the types of young meadow found on my land, and in many abandoned farm fields and other transient, sunny places. In many cases, the foundational species may be the same — little bluestem or Indian grass among the bunchgrasses, a host of asters and goldenrods among the dominant herbs, with various milkweeds and mountain mints throughout. In the list below, you'll find grassland specialists as well as some generalists that thrive in sunny disturbed places. You'll also find some shrubs that establish in grasslands and can be maintained as a part of a sunny, open mosaic.

Sometimes meadows are the most practical restoration goal for a site. Forests take a long time, probably longer than a single human lifespan, to express their diverse potential. Meadows can be established relatively rapidly in a way that supports diverse plant species and abundant pollinators, as well as a variety of songbirds and mammals depending on the size of the opening and the presence or absence of native shrubs. Faced with a lawn or abandoned farm field and the choice of restoring it to meadow or forest, I'd choose meadow the vast majority of times. Even if forest is the eventual restoration goal, a period in meadow plants mirrors successional patterns and probably helps the transition from bereft soils to those that may eventually support a diverse forest community.

Indicator Species: Herbaceous Plants.
Dandelion,* plaintain,* Queen Anne's lace,* chicory,* horseweed, yarrow, various goldenrods, dogbane, purpletop, daisy fleabane, common milkweed, Indian tobacco, evening primrose.

* *Indicates non-native species.*

Table 8.5: Cultural Species of Meadows and Grasslands

Common	Latin	Form	Primary Use
Eastern Yarrow	*Achillea gracilis*	Herb	Medicinal tops
Redroot Amaranth	*Amaranthus retroflexus*	Herb	Edible foliage and seeds
Pearly Everlasting	*Anaphalis margaritacea*	Herb	Medicinal tops
Spreading Dogbane	*Apocynum androsaemifolium*	Herb	Cordage
Indian Hemp	*Apocynum cannabinum*	Herb	Cordage
Common Milkweed	*Asclepias syriaca*	Herb	Edible tops
Butterfly Milkweed	*Asclepias tuberosa*	Herb	Medicinal root
Goosefoot	*Chenopodium berlandieri* and spp.	Herb	Edible foliage
Spring Beauty	*Claytonia virginica*	Herb	All parts edible
American Hazelnut	*Corylus americana*	Shrub	Edible nuts
Hawthorn	*Crataegus* spp.	Small Tree	Medicinal flowers and fruits
American Persimmon	*Diospyros virginiana*	Tree	Edible fruit
Wild Strawberry	*Fragaria virginiana*	Herb	Edible fruit
American Pennyroyal	*Hedeoma pulegioides*	Herb	Tea from foliage
Annual Sunflower	*Helianthus annuus*	Herb	Edible seeds
Wild Bergamot	*Monarda fistulosa*	Herb	Medicinal foliage and flowers
Dotted Horsemint	*Monarda punctata*	Herb	Medicinal foliage and flowers
Evening Primrose	*Oenothera biennis*	Herb	Nutritious seeds
Pokeweed	*Phytolacca americana*	Herb	Edible shoots
American Heal-All	*Prunella vulgaris* var. *lanceolata*	Herb	Medicinal tops
American Plum	*Prunus americana*	Shrub	Edible fruit
Black Cherry	*Prunus serotina*	Tree	Medicinal bark
Chokecherry	*Prunus virginiana*	Shrub	Edible fruit
Rabbit Tobacco	*Pseudognaphalium obtusifolium*	Herb	Medicinal tops
Broadleaf Mountain Mint	*Pycnanthemum muticum*	Herb	Tea from foliage
Narrowleaf Mountain Mint	*Pycnanthemum tenuifolium*	Herb	Tea from foliage

Table 8.5: Cultural Species of Meadows and Grasslands (*Continued*)

Common	Latin	Form	Primary Use
Virginia Mountain Mint	*Pycnanthemum virginianum*	Herb	Tea from foliage
Smooth Sumac	*Rhus glabra*	Shrub	Beverage from seeds
Staghorn Sumac	*Rhus typhina*	Shrub	Beverage from seeds
Black Locust	*Robinia pseudoacacia*	Tree	Edible flowers
Allegheny Blackberry	*Rubus allegheniensis*	Shrub	Edible fruit
Northern Dewberry	*Rubus flagellaris*	Shrub	Edible fruit
Swamp Dewberry	*Rubus hispidus*	Shrub	Edible fruit
Red Raspberry	*Rubus idaeus* var. *strigosus*	Shrub	Edible fruit
Blackcap Raspberry	*Rubus occidentalis*	Shrub	Edible fruit
Sassafras	*Sassafras albidum*	Tree	Root for beverage
Common Blue Violet	*Viola sororia*	Herb	Edible foliage

PLANT COMMUNITY: FORESTED WETLANDS

Swamps can be challenging habitats to walk through. Some are so dense with shrubs we'll need to crawl in places. In others, standing water is deep and we'll hop from hummock to hummock, clinging to tree trunks for extra balance. Some prefer to wear high boots in these habitats, but once overtopped by a misstep or an unexpected patch of deep water, these become a burden. We may not be perfectly adapted swamp creatures, but few habitats match the allure of the swamp, where dark waters reflect the trees overhead and each stride can require careful forethought.

There's a tough crawl up ahead, underneath a thicket of sweet pepperbush, swamp azalea, mountain laurel, mountain holly, great laurel, and highbush blueberry. You'll have to get way down, but if you literally crawl your knees will sink deep in the sphagnum mosses and come up wet to the bone. Yes, that's a big pile of bear scat over there, don't put your knees in that either.

The shaded wetlands we were in earlier today were a lot easier to traverse. We were underneath yellow birch, red maple, and tuliptrees, and shrubby spicebush and serviceberry were present, but not very dense. Instead, a lush carpet of herbs grew on the spongy ground, hydrated by mineral-rich seeps and a broad, level stream corridor. There was a huge spread of dwarf ginseng, marsh blue violet, marsh marigold and wood anemone flowering just inches above the delicate fern moss and sphagnum, and a dense colony of wild leeks as well. Easy street compared to this shrub swamp!

But there's a reward here just ahead. See the conifers poking above the shrubs ahead? That's a ring of stunted black spruce growing along a sunny edge. Push aside this last swamp azalea, and ... breathtaking! We're out of the dark branchy thicket and squinting in bright sunshine across a bog mat carpeted in green, yellow, white, and red sphagnum mosses. Among them are the carnivorous plants roundleaf sundew and purple pitcher plant, as well as a scattering of leatherleaf and sheep laurel. I remember the day I found this place with my colleague Kerry. It was the best botanical fieldwork day of my life.

The word "swamp" is often used interchangeably with "marsh" and sometimes "bog." A swamp is a wetland habitat that is forested. In *intact* forested wetlands, shaded streams and swamps can braid, converge, and broaden, transitioning freely between the vividly flowing, clearly demarcated "stream" and the saturated expanses we call "swamps." Some wetlands will be fed by mineral-rich groundwater percolating through rich geology, but even in areas of nutrient-poor rock, the layering of organic material and the binding of nutrients in silts and clays can maintain high fertility compared to nearby uplands.

Plants of forested wetlands overlap significantly with those found in riparian areas (described above in *Riparian Forests*). However, in wetlands, the presence of water can be nearly constant, whereas the riparian habitat described above is a well-drained upland that may occasionally be inundated in flood events.

Indicator Species: Woody Plants. Red maple, tupelo, green ash, black ash, swamp white oak, pin oak, hemlock, arrowwood viburnum, swamp azalea, sweet pepperbush, highbush blueberry, Japanese barberry.*

Indicator Species: Herbaceous Plants. Cinnamon fern, skunk cabbage, false hellebore, fringed sedge, fowl manna grass, sensitive fern, poison ivy, New York fern, jewelweed, jack-in-the-pulpit, false nettle, Japanese stiltgrass.*

* *Indicates non-native species.*

Table 8.6: Cultural Species of Forested Wetlands

Common	Latin	Form	Primary Use
Red Maple	Acer rubrum	Tree	Syrup
Groundnut	Apios americana	Vine	Edible tuber
Black Chokeberry	Aronia melanocarpa	Shrub	Edible fruit
River Cane	Arundinaria gigantea	Grass	Craft/utilitarian materials
Pawpaw	Asimina triloba	Small Tree	Edible fruit
Marsh Marigold	Caltha palustris	Herb	Edible foliage
Spring Cress	Cardamine bulbosa	Herb	Edible foliage
Turtlehead	Chelone glabra	Herb	Medicinal tops
Blue Huckleberry	Gaylussacia frondosa	Shrub	Edible fruit
Sessile Bugleweed	Lycopus amplectens	Herb	Edible roots
Northern Bugleweed	Lycopus uniflorus	Herb	Edible roots
Tupelo	Nyssa sylvatica	Tree	Edible fruit
Swamp White Oak	Quercus bicolor	Tree	Edible nuts
American Black Currant	Ribes americanum	Shrub	Edible fruit
Swamp Saxifrage	Micranthes pensylvanica = Saxifraga pensylvanica	Herb	Edible foliage
Mad Dog Skullcap	Scutellaria lateriflora	Herb	Medicinal tops
Water Parsnip	Sium suave	Herb	Edible root
Highbush Blueberry	Vaccinium corymbosum	Shrub	Edible fruit

PLANT COMMUNITY: SUNNY WETLANDS AND SHORES

Whap-splash!!! What was that noise? Something just hit the water. If we're in our kayaks, and a bear swims up, what exactly happens?

Whap!!

We weave our kayaks through the shallow water. Sometimes there is barely enough room between hummocks and semi-submerged mucks to navigate. I back into a sedgy hummock with blue vervain, swamp rose, and lurid sedge growing on it. Then I wiggle the kayak forward, lodging into a shrubby mass of elderberry, meadowsweet, and climbing boneset.

Looking along the marshy pond's edge, we see a clue to the source of that alarming sound. All the trees are cut down, with jagged chiseled stumps jutting from the shore line. Thickets of swamp azalea, winterberry holly, silky dogwood, and highbush blueberry are thriving in the full sun and moist soils, with spotted Joe Pye weed in full bloom and bottle gentians in bud.

Those jagged stumps? They are the work of beavers, harvesting fresh wood for their winter stores. That *whap-splash* noise? The warning call of these marsh dwellers, commenting on our clumsy presence.

Marshes, pond edges, and wet meadows are sunny habitats permeated by water. In many cases, a gradient of wetness is found, from open water to drier hummocks and banks, and this allows for a diversity of plant species as well as plant types — aquatic, emergent, and terrestrial. Given the high availability of water and sun, these habitats can be very productive of food and fiber, and are often prolific in blooming, with vivid swards of Joe Pye weed, boneset, blue vervain, ironweed, swamp milkweed, swamp rose, and the like.

Marsh habitats often occur around slow-moving water. Silts settle here and broad marshy plains can develop. Some marshes result from streams that have been dammed by beavers. Water radiates out to flood, replenish, and then drain over time.

Historically beavers were very numerous in the Northeast. Not only did beaver action slow and spread water, but abandoned beaver ponds featured rich soils full of decomposing plant debris and retentive of nutrients such as nitrogen, calcium, phosphorus, and magnesium.[33] Beaver activity supports stream health: "In a healthy, beaver-rich stream, dams and ponds act as step-like gradient controls that slow down flows and alleviate erosion," reports Ben Goldfarb, author of *Eager: The Surprising Secret Lives of Beavers and Why They Matter.* Without beavers, streams can rush through their channels, incising deeper and deeper until they become devegetated gullies subject to erosion and collapse. Many former marshes and wet meadows are cut off from the water that once washed over and pervaded the sponge-like earth.

Worldwide, wetland habitats are estimated to have been reduced by 54–57% [34] due to widespread drainage practices for agriculture and urban development. Wetlands are among the world's most productive habitats, and this means they have a central role in sequestering carbon. Wetlands restoration is therefore a core strategy in drawing down atmospheric carbon, and a simple answer to this need may be to bring back beavers, who excel at impounding flowing water and creating hydrologically imbued landscapes.

With beavers presently less common, plants of sunny wetlands find homes in a variety of human-created habitats, included dug and dammed pond shores, wet meadows growing on former farm fields over clay where plowing created an impermeable hardpan, and even drainage infrastructure astride roads and other developed landscapes.

Indicator Species: Woody Plants. Red maple, pin oak, silky dogwood, swamp rose, buttonbush, winterberry holly, highbush blueberry.

Indicator Species: Herbaceous Plants. Japanese stiltgrass,* purple loosestrife,* woolgrass, bur-reed, jewelweed, spatterdock, white pond lily, common reed, coontail, tussock sedge, spotted Joe Pye weed.

* *Indicates non-native species.*

Table 8.7: Cultural Species of Sunny Wetlands and Shores

Common	Latin	Form	Primary Use
Red Maple	*Acer rubrum*	Tree	Syrup
Sweetflag	*Acorus americanus*	Herb	Medicinal foliage and rhizome
Purplestem Angelica	*Angelica atropurpurea*	Herb	Medicinal root
Groundnut	*Apios americana*	Vine	Edible tuber
Turtlehead	*Chelone glabra*	Herb	Medicinal tops
Silky Dogwood	*Swida amoma = Cornus amomum*	Shrub	Craft/utilitarian materials
Boneset	*Eupatorium perfoliatum*	Herb	Medicinal tops
Hollow Stem Joe Pye	*Eutrochium fistulosum*	Herb	Medicinal root
Spotted Joe Pye	*Eutrochium maculatum*	Herb	Medicinal root
Spotted St. John's Wort	*Hypericum punctatum*	Herb	Medicinal tops
Sessile Bugleweed	*Lycopus amplectens*	Herb	Edible roots
Northern Bugleweed	*Lycopus uniflorus*	Herb	Edible roots
American Lotus	*Nelumbo lutea*	Aquatic Herb	Edible flowers, shoots, roots, and young seeds
White Pond Lily	*Nymphaea odorata*	Aquatic Herb	Edible flower buds
Pickerelweed	*Pontederia cordata*	Aquatic Herb	Edible seeds
Narrowleaf Mountain Mint	*Pycnanthemum tenuifolium*	Herb	Tea from foliage
Broadleaf Arrowhead	*Sagittaria latifolia*	Aquatic Herb	Edible tubers
Black Willow	*Salix nigra*	Tree	Medicinal bark
Elderberry	*Sambucus canadensis*	Shrub	Edible fruit and flowers
Gamma Grass	*Tripsacum dactyloides*	Grass	Edible seeds

Table 8.7: Cultural Species of Sunny Wetlands and Shores (*Continued*)

Common	Latin	Form	Primary Use
Broadleaf Cattail	*Typha latifolia*	Herb	Edible pollen and stems
Blue Vervain	*Verbena hastata*	Herb	Medicinal tops
Wild Rice	*Zizania aquatica*	Grass	Edible seeds

PLANT COMMUNITY: HIGH ELEVATION AND NORTHERN FORESTS

We're going to have to travel for this one. Not that cold swamps in northern New Jersey don't have some of these species, but many of them are rarities. High elevation? The tallest mountain in my state tops out at 1,804 feet (people from out west, please don't laugh). So ... we could drive south along the Appalachian chain and pick up some high peak habitats in the Smokies, or we could go way up into Canada and check out the Gaspé Peninsula, it's been on my wish list for a long time. But let's keep our drive to a sane one day's worth. I'll take us to one of the most spectacular places on the Atlantic Coast, in Downeast Maine.

We park on some gravel and walk through a thickly mossed forest of spruce, fir, and yellow birch. Under the deep green shade of the conifers, the fern moss and feather moss glow gold and green, punctuated by the occasional bright white of reindeer lichen and the red fruits of bunchberry. Everything is so vivid. We walk through a thick sward of wild sarsaparilla in fruit, and taste the sweet, currant-like fruits with that odd soapy finish typical of Aralia. Further on, a beautiful sward of bluebead lily and rose twisted-stalk presage an even more dramatic sight: we've reached the ocean. Looking out from a chasm incised into the sea-worn cliffs, we discover that we're high on a granite bluff, with waves crashing in violent white breakers down below. We stop to identify a few plants: roseroot growing in a cliff crevice, and bristly gooseberry clinging to a sheer slope. The gooseberry is delicious, not so the aptly named skunk currant we find next as we continue through the forest. Don't worry. We'll pig out in a few minutes, when we reach the berry glades. Growing on thinner soils on top of the granite cliffs, these host lowbush blueberry, velvetleaf blueberry, mountain cranberry, black chokeberry, chokecherry, pin cherry, bunchberry, and more. Welcome to the north!

With a short growing season and harsh conditions, the species of northern and high elevation forests are a select lot. Also included in this grouping are boreal forests like the one we just visited in Maine. I will treat these three related forest types — northern, boreal, and high elevation — together here due to their many overlapping species. They are farther away from where I live, and if I took out the knife to start cutting them up, I'm afraid I'd make a mess of it. My community description is broad and includes a number of different habitat types, sunny and shaded, moist and dry, just as our walk above did. Unlike some of the community types described earlier, you'll find species from wetlands to uplands in the list below, but all with an affinity for these northern and high-elevation forests. Further research on your part is encouraged so you can aim for fidelity to your local conditions.

These forests are often characterized by conifers in the canopy, including the spruces and firs of the boreal forest. In places with less cold conditions the classic association leans towards beech, sugar maple, basswood, and yellow birch, with hemlocks and white pines representing the evergreen set.

Deciduous species have less of an advantage up north (and up high in the mountains) as it becomes more expensive for deciduous species to lose and regrow foliage every year, given the short growing season. Still, boreal forests are lush, abundant places, often with thick carpets of moss harboring low herbs and a fragrant canopy of spruce, fir, northern white cedar, hemlock, or white pine. Birches, maples, beech, aspen, and other deciduous trees are also core members of these forests.

Northern forests were dramatically shaped by glaciers, which carved out river systems, reshaped mountains, and left water-filled depressions in their wake. Upland soils are often comprised of coarse glacial till, with hilltops and summits shaved to the bedrock. Meanwhile, low, flat areas can feature clay deposits that retain moisture at the soil surface and host wetlands.

Indicator Species: Woody Plants. White pine, white spruce, black spruce, red spruce, eastern hemlock, paper birch, yellow birch, quaking aspen, beech, basswood, sugar maple, striped maple, hobblebush, witherod, American red raspberry, steeplebush, bunchberry.

Indicator Species: Herbaceous Plants: Whorled wood aster, bluebead lily, fireweed, wild sarsaparilla, goldthread, starflower.

Table 8.8: Cultural Species of High Elevation and Northern Forests

Common	Latin	Form	Primary Use
Balsam Fir	*Abies balsamea*	Tree	Medicinal resin and foliage
Red Maple	*Acer rubrum*	Tree	Syrup
Mountain Serviceberry	*Amelanchier bartramiana*	Shrub	Edible fruit
Wild Sarsaparilla	*Aralia nudicaulis*	Herb	Medicinal root
Yellow Birch	*Betula alleghaniensis*	Tree	Bark tea
Paper Birch	*Betula papyrifera*	Tree	Bark for containers
Blue Cohosh	*Caulophyllum thalictroides*	Herb	Medicinal root
Fireweed	*Chamaenerion angustifolium = Epilobium angustifolium*	Herb	Edible shoots
Goldthread	*Coptis trifolia*	Herb	Medicinal rhizomes
Beaked Hazelnut	*Corylus cornuta*	Shrub	Edible nuts
Hawthorn	*Crataegus* spp.	Small Tree	Medicinal flowers and fruits
American Beech	*Fagus grandifolia*	Tree	Edible nuts
Creeping Wintergreen	*Gaultheria hispidula*	Subshrub	Edible fruit and foliage
Chocolate Root	*Geum rivale*	Herb	Root for beverage
Common Juniper	*Juniperus communis*	Tree	Flavoring agent
Tamarack	*Larix laricina*	Tree	Medicinal bark, edible young foliage
Ostrich Fern	*Matteuccia struthiopteris*	Fern	Edible shoots
Indian Cucumber Root	*Medeola virginiana*	Herb	Edible rhizome
Sweetgale	*Myrica gale*	Shrub	Tea and culinary herb
White Spruce	*Picea glauca*	Tree	Edible tips, medicinal resins
Black Spruce	*Picea mariana*	Tree	Edible tips, medicinal resins
Red Spruce	*Picea rubens*	Tree	Edible tips, medicinal resins
Jack Pine	*Pinus banksiana*	Tree	Medicinal foliage, resin, and pollen
Red Pine	*Pinus resinosa*	Tree	Medicinal foliage, resin, and pollen

Table 8.8: Cultural Species of High Elevation and Northern Forests (*Continued*)

Common	Latin	Form	Primary Use
White Pine	*Pinus strobus*	Tree	Medicinal foliage, resin, and pollen
Balsam Poplar	*Populus balsamifera*	Tree	Medicinal buds
Chokecherry	*Prunus virginiana*	Shrub	Edible fruit
Roseroot	*Rhodiola rosea*	Herb	Medicinal root
American Gooseberry	*Ribes hirtellum*	Shrub	Edible fruit
Red Raspberry	*Rubus idaeus var. strigosus*	Shrub	Edible fruit
Purple Flowering Raspberry	*Rubus odoratus*	Shrub	Edible fruit
Northern White Cedar	*Thuja occidentalis*	Tree	Smudge, tea, cordage
Basswood	*Tilia americana*	Tree	Cordage from bark
Eastern Hemlock	*Tsuga canadensis*	Tree	Tea from foliage
Lowbush Blueberry	*Vaccinium angustifolium*	Shrub	Edible fruit
Cranberry, Lingonberry	*Vaccinium macrocarpon, Vaccinium vitis-ideae*	Subshrub	Edible fruit
Velvetleaf Blueberry	*Vaccinium myrtilloides*	Shrub	Edible fruit
Lowbush Blueberry	*Vaccinium pallidum*	Shrub	Edible fruit
Hobblebush	*Viburnum lantanoides*	Shrub	Edible fruit
Witherod	*Viburnum nudum*	Shrub	Edible fruit
Highbush Cranberry	*Viburnum trilobum*	Shrub	Edible fruit

PLANT COMMUNITY: SANDY PINE AND OAK FORESTS

"*In the pines, in the pines, where the sun, never shines ...*" I can't help but sing the old Leadbelly tune to myself when we head to the New Jersey Pine Barrens. But the truth is that when I head to the Pine Barrens, I usually choose to walk along sunny edges and openings. Don't get me wrong. I love walking in the shade of shaggy pitch pines, instantly recognizable by the needle clusters sprouting here and there along trunk and branches. I love smelling the volatilizing pine terpenes coming up from the duff on a hot day. But, although the understory beneath the pines and oaks is often dominated by heaths such as black huckleberry and lowbush blueberry, the best place to forage fruits and find wildflowers is along the sand roads and dikes of former cranberry bogs.

We bump along narrow roads dividing the inundated ponds that used to be cranberry-production bogs. We're taking my old green 4Runner. I'm happy to have four-wheel drive here, though it won't save us from some of the pits where the road has slumped into the tannic water below. On your side of the vehicle we're passing an older wetland now, an Atlantic white cedar swamp with sweetbay magnolia on its edges. Its red fruits are showing along with winterberry holly.

Up ahead is a spot where a colleague directed me, a place to see a very rare native morning glory growing in the exposed sands along the road. If you come back here around July 4th, you can harvest a practically infinite supply of highbush blueberry, black huckleberry, lowbush blueberry, and dangleberry along the overgrown dikes that run transversely across the old bogs.

Pinelands and other pine- and oak-dominated coastal plain communities are found over the remains of ancient coastlines when sea level was higher than it now is. They retain some of the topography of the maritime strand, with ancient paleodunes forming uplands. They are often nutrient poor, over excessively drained sands, except where lenses of clay have formed in bottomlands where fine

particles have settled and create poorly drained soils.

The plant communities of coastal pine and oak forests overlap with those found along the present-day seacoast (see below), as well as those found in acidic upland forests, especially in sandy and gravelly glaciated soils (see above).

Forests on well-drained sands are fire-prone habitats. Many of the dominant species are fire-adapted in some way. Pitch pine, for example, has serotinous cones that open to shed their seeds after intense heat; it also features adventitious buds along its trunk that can resprout if fire kills the tree's leader. Likewise, many of the heath species such as huckleberries and blueberries sprout readily after fire events.

While the abandoned cranberry fields we visited were artificial in origin and structure, seeking out sunny spots in the Pine Barrens makes a lot of sense. Fire acts as a canopy-thinning agent in the Pine Barrens and encourages diversity in the ground layer, and many of the characteristic species of the pinelands habitat are dependent on sunny conditions, whether they arise from controlled burns or are found along sand roads.

The same is true of many of the sandy forests that extend along the coastal plain from Massachusetts all the way into Florida. While the pine species vary, from pitch pine and Virginia pine to shortleaf, loblolly, and longleaf pines, the entire system of dry pinelands is in many ways dependent on fire. Fire ensures successful reproduction of the pines themselves, and maintains the diversity of herbs and grasses in the understory. Where fire is less frequent, oaks and other hardwoods dominate, creating their own characteristic forests but without the distinctive aroma of warm pine needles on a sunny day.

Indicator Species: Woody Plants. Pitch pine, Virginia pine, American holly, white oak, sweetgum, tupelo, shrub oak, sweet pepperbush, sweetfern, black huckleberry, lowbush blueberry, yaupon.

Indicator Species: Herbaceous Plants. Pennsylvania sedge, switchgrass, broomsedge, little bluestem, bracken fern.

Table 8.9: Cultural Species of Sandy Pine and Oak Forests

Common	Latin	Form	Primary Use
Red Maple	*Acer rubrum*	Tree	Syrup
Black Chokeberry	*Aronia melanocarpa*	Shrub	Edible fruit
Pawpaw	*Asimina triloba*	Small Tree	Edible fruit
Allegheny Chinquapin	*Castanea pumila*	Shrub	Edible nuts
Sweetfern	*Comptonia peregrina*	Shrub	Tea
One-flowered Hawthorn	*Crataegus uniflora*	Shrub	Medicinal flowers and fruits
American Persimmon	*Diospyros virginiana*	Tree	Edible fruit
Creeping Wintergreen	*Gaultheria hispidula*	Subshrub	Edible fruit and foliage
Black Huckleberry	*Gaylussacia baccata*	Shrub	Edible fruit
Blue Huckleberry	*Gaylussacia frondosa*	Shrub	Edible fruit
Sweetgrass	*Hierochloe odorata*	Grass	Craft/utilitarian materials
Yaupon	*Ilex vomitoria*	Shrub	Tea from foliage
Sweetgum	*Liquidambar styraciflua*	Tree	Medicinal resin
Dotted Horsemint	*Monarda punctata*	Herb	Medicinal foliage and flowers
Bayberry	*Morella pensylvanica*	Shrub	Tea from foliage
Tupelo	*Nyssa sylvatica*	Tree	Edible fruit
Maypops, Purple Passion Flower	*Passiflora incarnata*	Vine	Edible fruit

Table 8.9: Cultural Species of Sandy Pine and Oak Forests (*Continued*)

Common	Latin	Form	Primary Use
Shortleaf Pine	*Pinus echinata*	Tree	Medicinal foliage, resin, and pollen
Pitch Pine	*Pinus rigida*	Tree	Medicinal foliage, resin, and pollen
Loblolly Pine	*Pinus taeda*	Tree	Medicinal foliage, resin, and pollen
Virginia Pine	*Pinus virginiana*	Tree	Medicinal foliage, resin, and pollen
Chickasaw Plum	*Prunus angustifolia*	Shrub	Edible fruit
White Oak	*Quercus alba*	Tree	Edible nuts
Northern Dewberry	*Rubus flagellaris*	Shrub	Edible fruit
Swamp Dewberry	*Rubus hispidus*	Shrub	Edible fruit
Sassafras	*Sassafras albidum*	Tree	Root for beverage
Blue Mountain Tea	*Solidago odora*	Herb	Tea from foliage and flowers
Highbush Blueberry	*Vaccinium corymbosum*	Shrub	Edible fruit
Lowbush Blueberry	*Vaccinium pallidum*	Shrub	Edible fruit

PLANT COMMUNITY: THE SEASHORE

Stand with me along the seacoast, the maritime strand of the coastal plain, windswept, wave-wracked. The wind is picking up and sand is needling our legs, while the waves transport unimaginable quantities of the stuff in the rolling surf. This barrier island, and the shifting lines of dunes, are so geologically ephemeral yet such a critical buffer. They're like skin shielding the body of the inland province, constantly shedding and regenerating.

We've stepped out onto the beach after collecting beach plums on the back dunes. These little plum shrubs appear all the shorter because their trunks and branches are constantly being buried a little deeper in sand. In our short walk to the beach, we pass from the back dunes with beach plums, seaside goldenrod, Virginia creeper and swards of beach heather, to the seaward-facing dunes with beachgrass and nary anything else, other than sea rocket along its edges.

Up and down the East Coast, the band where land and sea intergrade has been heavily developed. The ecological function of natural dune barriers is diminished, and storms and waves gnaw at our "castles" built on the sands, the sprawling disturbia of resort villas, bungalows, and particle-board mansions.

Billions of tons of terrestrial rock are dissolved each year, much of this largesse reaching the ocean eventually and contributing many elements and massive amounts of sediment.[35] This flow is then reversed in some small part as the ocean delivers sand and salt back to the land in that narrow interface most of us call the beach.

Just back of the beach — the foreshore and beach berm, that is — begin the dunes. These are sculpted by wind action and waves, but stabilized by plants. Beach grass (*Ammophila breviligulata*) is the primary actor in stabilizing the youngest seaward dunes, preparing them for colonization by other species.

Salt spray is a considerable factor in shaping dune plant communities. Beach grass is exceptionally tolerant of salt. The primary dunes closest to the waves are in a direct line to receive saline sprays, partially shielding secondary dunes leeward of the ocean. Salt tolerance is a powerful filter for what will and won't grow, and where.

Somewhat more protected, back dunes often harbor a thicket of low-growing woody plants, grasses, and some herbs. Among the woody plants of the thickets are numerous species bearing choice fruits,

including beach plums, serviceberry, *Aronia* species, shrub-like black cherry trees, and more. Bayberry is a prominent, aromatic member of the thicket community, growing in deep green rounded mounds. Grasses include several *Panicum* species such as switchgrass and beach panicgrass, as well as little bluestem. Herbaceous diversity increases away from the primary dunes as well.

Extreme sunbaked areas form on lenses of hot sand interspersed among the thickets. Low beach heather forms a stubbly blanket in these areas, and anyone walking through feels the crunch of reindeer lichen underfoot.

Low forests anchor the most durable dunes, with trees absorbing some of the winds that shape and reshape dunes. Windswept craggy-crowned American holly, red cedar, and several oak and pine species grow in these dense maritime forests.

Vines and creepers thrive throughout, well adapted to the ever-shifting landscape and the annual peregrinations of sun and shade.

Relatively modest changes in topography can have strong effects on plant assemblages. Slight depressions in the landscape accumulate clayey muds and retain freshwater from rains. Pocket marshes contain a diversity of sedges and rushes, as well as tall forbs such as swamp rose mallow, the native hibiscus. On the margins of these wet areas are found highbush blueberry, arrowwood viburnum, sweet pepperbush and other woody plants.

Along the extensively developed northeastern seaboard, vacation homes and other development have replaced natural communities in many areas. The dune system is a critical element of the interplay of land and sea, and its disappearance leads to the loss of beach sand (which is replaced from dunes) and to radically increased risk of flood damage during storm events.

American beachgrass (*Ammophila breviligulata*), is an extremely important beach restoration species. It grows with increased vigor when buried, traps sand in its filter-like mesh of upright leaf blades, and spreads rapidly via rhizomes. These characteristics make it singularly effective at building dunes — hence, it is a pioneer species that can be said to create its own habitat. *Ammophila* also benefits from association with mycorrhizal fungi that provide enhanced nutrient and water availability to the plant in return for sugars, the product of photosynthesis. American beachgrass loses vigor when not in an area of ongoing sand accumulation, due to diseases and herbivory, including a fungal blight.[36] It can function as a short-lived pioneer species in other areas, providing shade for the germination of other species.[37] While not an edible or medicinal species, this is such a core species in its niche that it is worth being aware of in any restoration involving coastal areas.

As one proceeds towards the southeast, another grass, sea oats, *Uniola paniculata*, plays an important role similar to *Ammophila*. It ranges from Virginia southward. The seeds of sea oats are edible when ground into flour (which is the case for grasses in general, but some are more worth the effort in terms of size and efficiency). It is not a prolific producer of viable seed.[38]

While initial primary dune restoration is usually accomplished with extensive plantings of beachgrass, more inland habitats can be restored with a diverse palette of edible and other functional species.

Similar communities are found along the shores of the Great Lakes, which also feature impressive dunescapes, though they lack the influence of salt water found along the seacoast.

Indicator Species: Woody Plants. American holly, red cedar, southern red oak, live oak, shining sumac, groundsel-bush, Virginia rose, rugosa rose,* beach heather, scrub oak.

Indicator Species: Herbaceous Plants. Yarrow, Virginia pepperweed, rabbit tobacco, seaside goldenrod, bracted plantain, little bluestem, Beach panic grass, Japanese sedge,* spotted knapweed,* Japanese brome,* sheep sorrel.*

* *Indicates non-native species.*

Table 8.10: Cultural Species of the Seashore

Common	Latin	Form	Primary Use
Canadian Serviceberry	*Amelanchier canadensis*	Small Tree	Edible fruit
Black Chokeberry	*Aronia melanocarpa*	Shrub	Edible fruit
American Persimmon	*Diospyros virginiana*	Tree	Edible fruit
Virginia Pepperweed	*Lepidium virginicum*	Herb	Edible foliage and seeds
Starry Solomon's Plume	*Maianthemum stellatum*	Herb	Edible shoots
Eastern Prickly Pear	*Opuntia humifusa, O. cespitosa*	Cactus	Edible fruit and pads
Beach Plum	*Prunus maritima*	Shrub	Edible fruit
Black Cherry	*Prunus serotina*	Tree	Medicinal bark
Rabbit Tobacco	*Pseudognaphalium obtusifolium*	Herb	Medicinal tops
Seaside Goldenrod	*Solidago sempervirens*	Herb	Tea from foliage and flowers
Highbush Blueberry	*Vaccinium corymbosum*	Shrub	Edible fruit
Lowbush Blueberry	*Vaccinium pallidum*	Shrub	Edible fruit
Scuppernong, Muscadine	*Vitis rotundifolia*	Liana	Edible fruit

CHAPTER 9

Plant Species

THE PURPOSE OF THE PLANT ACCOUNTS below is to entice you with the personality and possibilities of native plant species that are edible, medicinal, used for crafts, or otherwise employed culturally. Each account presents a portrait of a native plant and its cultural uses. If I have a good, brief story about a plant I tell it. Habitat information is included to help you in selecting appropriate species for appropriate places, and can be used in tandem with the Plant Community descriptions in the chapter above.

These accounts are not comprehensive. Each plant here could be written about in an extensive monograph, or studied for a lifetime. *These profiles don't guarantee safety or efficacy of edible or medicinal uses.* They are invitations for you to learn more. Oftentimes safety and efficacy are not absolutes but are based on proper or improper usage or dosage. These are good things to learn through experience and careful research.

I have included historical and contemporary plant usages by Indigenous people whenever possible. Where I have been general or have paraphrased, I hope I have not disrespected people's cultural traditions. The same goes for any simplifications I have made of the practices of the various contemporary herbalists cited below.

Mentions from early colonial documents of the intentional cultivation of a variety of native species by Indigenous peoples are of particular interest and relevance. I have included them to add historical depth to our picture of the utilization and management of the native species of this continent. People have done this before and we can do it again!

When I first became interested in herbal medicine, one of the things that perplexed me most was the multitude of disparate-seeming uses for a single herb. I think I expected something more like the specificity of an over-the-counter cough medicine or pain reliever. How could I pick among the many overlapping or contra-dictory-seeming uses of so many herbs? How could one herb be good for diarrhea, varicose veins, bleedings gums, and poor digestion? In this case, it is because the herb in question, white oak bark, is an astringent. It tones and dries body tissues that are swollen, lax, or leaky, through the physical properties of tannins, which bind proteins. Astringency is what is known as an herbal "action," as is being sedative, or antispasmodic, or stimulant. Herbs un-derstood using these categories will have actions that can often be useful across body systems, though witch hazel and stoneroot might be better astringents for hemorrhoids, and blackberry root might be better for diarrhea. I initially learned about herbal actions (and energetics) primarily through the writings of Matthew Wood, and interviews I listened to with Jim McDonald, and I recommend both their work whole-heartedly.

The plant profiles below may include visual descriptions but are not generally intended to help in identification. For that, I'd recommend a good field guide or two, like *Newcomb's Wildflower Guide*. You could rely on an app to do your ID for you, but then you miss out on the good part. Do you want a robot to eat your chocolate ice cream while you read the label?

Knowing a plant and a thing or two about its edible uses is only half the battle, as they say. Here, eat this raw potato! Not

so good, huh? Preparation is key. As are harvesting and processing methods. A good wild foraging guide will provide you with all those details, and there are no finer books on the topic than Sam Thayer's books *The Forager's Harvest, Nature's Garden,* and *Incredible Wild Edibles.* Thayer is an astute botanist, an incredibly thor-ough investigator, and a beautiful and at times very funny writer. If you don't know these books already you're in for a treat.

I occasionally discuss the ranges of the plants below in general terms, but I strong-ly encourage you to find up-to-date range maps which show the states, provinces, and even counties where each species below is native, as part of your decision-making process regarding introducing a plant spe-cies to your restoration site.

Here are some plants. Fall in love!

Abies balsamea • Balsam Fir
Fragrant conifer is a medicine chest.
If you step into our house in the winter and smell a delicious rich smell (as opposed to fermenting kimchi or leftover chainsaw fumes on old jeans), chances are you're en-joying balsam fir. We often simmer balsam fir foliage on the woodstove or stovetop at home. It has a very rich, deep, and soothing aroma, and it's a pleasure to walk into the home in winter and be greeted by its warm-ing fragrance.

Balsam fir is a common species of both uplands and wetlands in the boreal region and at the highest elevations in the Appalachians. This aromatic, terpene-rich conifer has myriad uses in the traditional pharmacopeia of boreal peoples. Many of these uses are being investigated and confirmed using lab analysis. Perhaps I should use quotes around "confirmed" because, ultimately, what's more compelling — centuries of use with actual human beings, or a few transient lab trials?

Balsam fir oleoresin, harvested from blisters on the bark of young and medium-age trees, is used by Indigenous people to treat infections. Recent research indicates the activity of its combined terpenes and resin acids against Gram-positive bacteria.[39] In another study confirming traditional usage, three compounds isolated from *Abies balsamea* were found to enhance glucose storage in liver cells and reduce glucose output in liver cells. This suggests applications in the treatment of insulin resistance and type 2 diabetes.[40] The sap of the tree is used to heal burns.

Anishinaabe author Mary Siisip Geniusz describes the proper method for preparing balsam fir tea thus: "Pour boiled, and slightly cooled, water over the needles. Allow the tea to steep, covered, for 1 minute but no longer." More extended steeping yields a tea that tastes "like turpentine," she states, but brewed correctly, it is light and flavorful. I find it to be almost fruity in flavor, with deep resinous bass notes.

Some growers are experimenting with coppicing balsam fir for Christmas tree sales. After the tree is cut to the base, multiple sprouts ensue, each of which can be harvested as a separate tree. The root system persists despite the cutting, and the site is more stable and suitable for companion forest species than if trees were simply killed and replaced with seedlings.[41]

Acer: The Maples

Sweet alternatives.

If you consider our sources of sugar — cane with its dependence on slavery and colonialism, corn with its extensive monocultures and herbicide dependency — it is clear that our sweet tooth has led to some highly problematic social and environmental consequences. By comparison to cane and corn, sugar maple grows as part of long-lived natural communities. Its sugar is rendered through nothing more complicated than a hole, a tap, and a vessel to boil the sap down in. If the environmental and human cost of cane sugar and corn syrup across the centuries were factored into the market price of those products, if the toll that processed sugar takes on human health were part of its price, maple sugar would be recognized for the incredibly good bargain that it is.

Red Maple (*Acer rubrum*): This ubiquitous tree species thrives in every single habitat discussed in this book. Red maple grows almost everywhere except under dense shading, from bald cypress-leather fern swamps in subtropical Florida to fens in the far north where it may be found with larch and cinnamon fern.[42] It is extremely frequent in uplands as well, often forming a subcanopy in oak forests where

fire is absent. Red maple is a common species in many forest types but typically a dominant species in wetlands, where it is abetted by its strong tolerance for waterlogged soils, and unhampered by the threat of fire.

As with other maples, red maple can be tapped for syrup. Sap sugar content estimates vary wildly depending on where studies are done, so I'll refrain from quoting statistics. However, red maple can sometimes come close to sugar maple in sap sugar content, and is widely used for maple syrup production in the south where sugar maple is out of range.[43]

There are some medicinal uses in the traditional repertoire. Herbalist Darryl Patton recounts his mentor, Tommie Bass, using red maple as a "slow but sure remedy for hot flashes." [44]

Silver Maple (*Acer saccharinum*): Silver maple lofts deeply-lobed foliage above spreading, often enormous trunks. It keeps company with other prodigious growers such as sycamore and tuliptree in riverine forest canopies. It has been shown to survive periods of inundation as long as one year,[45] but generally favors moist, draining soils with intermittent, not continuous, flood conditions. Like other riparian trees such as box elder and the aforementioned sycamore, silver maple can invade other, human-disturbed habitats, that may mirror flood-scoured river edges in offering bare soils and low

Red maple (*Acer rubrum*).

competition. It does not regenerate in shaded conditions.

This fast-growing tree supplies a sweet sap that can be converted into maple syrup or sugar. While not quite as high in sugar content as the sap of a sugar maple (though studies vary wildly), it makes a good substitute in habitats exposed to flooding and inundation, or those too hot for sugar maple. The period for tapping is roughly the same — late winter when daytime temperatures are above freezing but nighttime temperatures are not. In my experience, the syrup (and even the sap, straight from the tree) is abundantly sweet, but does not have the richness of flavor possessed by sugar maple.

Its wood is hard but brittle and silver maple has a reputation for easily shedding branches. Thus, it should be sited with care, away from structures. Mature specimens can be truly enormous in size.

Sugar Maple (*Acer saccharum*): Sugar maple thrives on slopes with deep, moist soils in the northern tier of the eastern forest. It has an affinity for rich soils and also creates them by generating a nutrient-rich leaf litter. It can be deeply shading when growing densely (as often happens with saplings in second growth forests) but as a mature, well-spaced tree it coexists with and supports a remarkable native understory of herbs and shrubs. It is moderately to highly shade tolerant depending on the quality of the site's soils.

Achillea gracilis • Eastern Yarrow
Many continents, many uses.

Found in a variety of dry open habitats from seaside dunes to high-elevation glades in the Appalachians, yarrow is a ubiquitous species of dry openings. There exists a range of opinions on the native status of this species; older sources designate our species as *Achillea millefolium*, which is an aggregate of races native to several continents. The current consensus is that both native and introduced species exist, but that the native species is most common in natural areas in eastern North America.[46]

Yarrow's native status is reinforced by the extensive use of this plant by Indigenous peoples in North America. It has immense cultural importance for numerous Indigenous peoples, with uses recorded for 58 different cultures in Daniel Moerman's ethnobotanical encyclopedia *Native American Medicinal Plants*.

Primarily used as a medicinal species, yarrow's uses include stopping bleeding, treatment of fever and colds, and for wounds and pain. Yarrow is a good example of the multitude of medicinal uses that can be found in one plant. One constituent, the alkaloid achileine, is known to enhance blood clotting.[47] Sesquiterpene lactones in yarrow have antimicrobial and anticancer activity. Constituents in the volatile oil are pain-relieving, expectorant, and diaphoretic (producing or increasing perspiration).[48] The unique combination of all these actions leads to the personality of the herb, held in high reverence in the Indigenous North American as well as European herbal traditions.

Acorus americanus • Sweetflag

Sword-leafed marsh dweller shines in the psychological realms.

This sword-bladed herb and its close relatives are known from the cultural and medical practices of several continents. Timothy Motley tells a great story about a band of thieves from medieval Europe using *Acorus calamus*, a close relative of the American species, in his monograph:

> During the time of the plague in Europe, when people avoided the sick and dead in order to avoid contracting the disease, a band of thieves was robbing the homes of infected people without being stricken by the disease. When the thieves were finally caught and brought to trial they were offered an acquittal if they would reveal the secret of their immunity. The mixture they drank and doused themselves with before their crimes came to be known as the "Vinegar of the Four Thieves." The potion contained beach and roman wormwood, rosemary, sage, mint, rue, lavender flowers, calamus powder, cinnamon, cloves, nutmeg, and garlic. The ingredients were placed in a crock and red wine vinegar poured over it, then left in the sun for several weeks and distilled off, followed by the addition of camphor. [49]

That's a potent cocktail of antibacterial herbs, and the inclusion of calamus (another name for sweetflag) points to its utility as an antibiotic agent, one of its numerous uses. Herbalist Jim McDonald elaborates: "Bitterroot [another alternative name for sweetflag] is strongly antimicrobial; chewing the root not only directly fights the infection (especially for throat colds), but is also stimulating and helps to overcome the rundown feeling that you get with a cold."

Anthropologist Melvin Gilmore, who authored a study of Indigenous peoples of the Missouri river region, observed that sweetflag was held in high esteem by all of the peoples he studied. Its uses included being a remedy for cough, fever, toothache, colic, and colds, and also ceremonially in garlands, "for their pleasant odor." Sweetflag has a wonderful aroma somewhat reminiscent of vanilla and citrus, with an underlying sweetness. Gilmore also relates that the plant was ascribed mystic powers, and that Teton Dakota warriors chewed calamus roots to prevent fear and to promote focus in battle.

We begin to see elements of sweetflag's personality emerge from these ethnobotanical notes — as a remedy for colds that also has ceremonial significance and can calm or focus the mind.

Gilmore was a fascinating proto-ethnobotanist. He noted that the range of sweetflag was closely correlated with old Pawnee village sites, and speculated that its spread was likely a result of human agency, whether through deliberate transport of its seeds or rhizomes, or by accidental spread, for example, in the use of its seeds as decorative beads. Gilmore broached that "the human factor in plant distribution prior to the European advent is not so obvious and may not have suggested itself to most of my readers. But the people of the resident tribes traveled extensively and received visitors from distant tribes. Their wants required for various purposes a great number of species of plants from mountain and plain and valley, from prairie and from woodland, from regions as remote from each other as the Rio Grande and the Great Lakes and St. Lawrence."

This type of far-ranging dispersal by human beings is core to the story of *Acorus americanus* and its relatives.

Until recently, our native species, *A. americanus*, was taxonomically lumped with the cosmopolitan species *Acorus calamus*, sometimes as *Acorus calamus* var. *americanus*.

Acorus calamus is a triploid species (having three sets of chromosomes), probably originating in Asia, which has been introduced in Europe, Australia, North America, South Africa, and elsewhere. It figures prominently in herbal usage in many of these places. It was supposedly introduced to Europe by invading Mongols, who reputedly planted it in every water source they found as they arrived in new territory.

Acorus americanus, our native, diploid species, is notable because it lacks a potentially carcinogenic phytochemical, beta-asarone, which is found in varying quantities in the triploid species. Extracts of the Asian calamus are banned in the United States due to potential toxicity, but *Acorus americanus* is nowhere specified as illegal, and lacks the problematic constituent. As an aside, beta-asarone may also be a mild hallucinogen.[50]

Our native sweetflag is differentiated from the triploid, non-native species by the presence of prominent secondary veins nearly equal in size to the midvein, as compared to *A. calamus* which has a single, prominent midvein. Other differences include that the Asian triploid species does not apparently produce viable seed, and that it is greater in stature (and more invasive) than the native plant. Note that there is still some disagreement among taxonomists.

Sweetflag has uses in the psychological realms. On his website at herbcraft. org, herbalist Jim McDonald reports that it reduces anxiety, helping errant thoughts to dissipate. He uses it for nervous stomach, and for anxiety attacks. McDonald also notes how it hones mental focus: "When preoccupations drop away, a clarity of perception is revealed. It becomes easy for the mind and senses to perceive; yet does so without narrowing our perception. Chewing on Calamus seems to sharpen vision noticeably, but ironically, this effect is more pronounced when you're just taking things in, and less so when you try to focus on something specific. When you do try and focus, the clarity diminishes somewhat; to return when again you return from 'looking' to 'seeing'." Other herbalists comment as well on its ability to restore focus to the mind, and Indigenous peoples of North America use the plant to restore the voice of singers. Some allude to its ability to induce an altered or heightened state of consciousness (see note above on beta-asarone).

I once chewed some sweetflag root (the triploid species, in this case). The herb communicated to me very clearly, and unbidden: *Don't use the word "work" for everything. Find another word for processes. Don't view everything through the lens of work, or you'll be working all the time.*

I'm working on it.

By the way, *High Times* magazine once reported that "[a]n ancient text from the Indian Ayurvedic tradition gives mention to smoking dried Calamus root with cannabis as an antidote that moderates [Cannabis] effects during medicinal use." [51]

In addition to its herbal uses, sweetflag can be candied as a confection. It is used to flavor beverages and in perfumes. Its leaves were strewn on the floors of homes and churches in days past, to deter pests and improve odors.

Acorus americanus spreads readily from rhizomes. Its preferred niche is in the shallow waters of marsh and lacustrine habitats.

Actaea racemosa • Black Cohosh
Deep woods medicine becomes international bestseller.

Black cohosh is a tall herb blooming in white spires at about face height, its several plateaus of horizontally aligned leaves broadening towards the base. Below ground lies a hard black rhizome replete with glycosides, phenolic acids, flavonoids, volatile oils, and tannins. [52] This long-lived forest herb has become one of the most popular herbal medicines of our time, deemed efficacious and safe [53] for menopausal symptoms.

The tens of millions of dollars worth of black cohosh extracts sold on the international market originate primarily from wild harvesting [54] and are problematic from a plant conservation perspective. Black cohosh thrives in high-quality forest habitats and is relatively slow to reproduce and mature.

One approach to supporting wild populations is to reproduce and grow this species ourselves. As with other wild medicinals, it is likely that the best quality medicinal plants will be grown in habitats most closely replicating the forests in which black cohosh is native. Black cohosh appreciates well-drained but moist soil, ample organic matter and soil nutrients, and some shade. Mature, well-established individuals are tolerant of full sun conditions such as occur in natural canopy gaps as well as utility right-of-ways and forestry operations. Reproduction from seed, however, occurs in shaded environments — this is not a meadow or glade species.

In addition to uses for menopausal discomfort, black cohosh has a history of use for musculoskeletal inflammation and finds its place in remedies for back pain, arthritis, and other related issues. It is also used by herbalists for certain types of depression. Recent pharmacological research suggests an influence on the central nervous system.[55]

Allium canadense • Wild Garlic

Pungent wild onion.

Wild garlic (AKA wild onion) grows in open woods on floodplains and river terraces, often in areas of greater light availability. Its flat basal leaves are overtopped or equaled by stiff stems bearing starry flowers and clusters of bulblets.

Wild garlic is very serviceable where an onion/garlic flavor is called for. Not as distinctive as ramps (see *Allium tricoccum*

Black cohosh (*Actaea racemosa*).

below), but much faster growing, it is regarded highly in terms of flavor. Given the diversity of wild onion relatives, it should be quite possible to reach allium self-sufficiency using a variety of perennial alliums, and this species can be at the center of that strategy. Wild onion can be enjoyed cooked, raw, or pickled. Prehistoric evidence for human consumption of wild onion dates as far back as 6,000 years, found in human coprolites in a Texas cave.[56] Porcher and Rayner, in their guide to wildflowers of South Carolina, suggest the following contemporary recipe: "Before flowering and after removing wilted leaves, the whole plant can be boiled in salted water and eaten as a delicious vegetable. The water is the base for a delicious cream-of-onion soup."[57]

As with other alliums, this species has a host of medicinal uses, including as a remedy for coughs, eye and ear infections, swellings, sores, stings, and clogged sinuses.[58]

Towards the Midwest, this species ranges into a variety of open prairie and disturbed habitats, whereas in the Northeast it generally has riparian affinities.

Allium cernuum • Nodding Onion

Perennial onion and beautiful bumblebee bloomer.

This native perennial onion ranges clear across the continent, growing on rocky ledges, in open upland woods, and in prairies. Its medium-sized bulbs are a staple food item for Northwest Indigenous peoples and are cooked in earthen pits and also used to flavor stews. The presence of inulin, a type of prebiotic, makes the bulbs sweet when cooked in pits (i.e. for a long time).[59] They can be used fresh or cooked

whenever an onion flavor is called for. The young foliage is also serviceable, though a bit coarse, where scallions or chives might be called for.

Nodding onion's name derives from its nodding umbel of showy flowers, which are very attractive to bumblebees.

Allium tricoccum • Wild Leek

Lush but fleeting seasonal delicacy in the deep woods.

Growing primarily on slopes and terraces above streams and rivers, this is a broad-leaved wild onion with a distinctive, delicious flavor to its foliage and bulbs. It is also known as "ramps" and has quite a following as such among foodies and rural folks alike. Its leaves arise in April and are generally dormant by June — an adaptation to forest life that it shares with other spring ephemerals such as trout lily and squirrel corn. It flowers in summer and produces seed by October.

This is a slow-growing species. It takes two years to germinate from seed, another two or three to reach a size ready for planting, with leaves maybe an inch thick. Reproductive individuals are likely seven years old or older, and the plant moves through the landscape via gravity-dispersed seed as well as bulb offsets. Thus, the huge colonies often encountered in its wild haunts are probably centuries old.

The recent faddishness of ramps as a gourmet food threatens its persistence in parts of its range, as large populations of humans can strip ancient colonies in relatively brief timespans. This recommends it for restoration on a broad scale. Humans are a good, highly mobile animal helper for ramps. Without help, the seeds may not move particularly quickly from one drainage to another.

A colleague has had considerable success introducing wild leeks via seed to the stream corridor woods on his property. Planting three year-old bulbs during dormant season would speed the process of colonization by at least a few years. However, given the timeframe for leek maturation and the commencement of reproduction, a few years here or there might not make much long-term difference.

Amelanchier: The Serviceberries

Lithe and elegant trees and shrubs with early fruit.

Amelanchier is a genus of small trees and shrubs, with individual species adapted to a wide variety of habitats from coastal dunes to mountain fens. They bear small fruit reminiscent of blueberries in size and shape.

Downy serviceberry (*Amelanchier arborea*), an elegant small tree, is a common understory species on steep, bouldery slopes and open summits. Like many trees, it has a wide breadth of habitat tolerances and could be utilized on most sites. Its profusion of graceful blooms are conspicuous

Wild leeks (*Allium tricoccum*).

in the early spring woods. They bloom prior to canopy leaf-out and the emergence of many herbaceous species.

I have many photos from botanical survey work through the years. On the first day of the field season, we're usually looking for spring ephemerals and a hint of the diversity to come. Every year I have at least a few blown-out pictures of serviceberries in full flower, their bright white flowers blazing against the still wintry grays of barren trunks with the crisp blue spring sky for background. You might guess from the

flowery language that I'm fond of this tree. Not only does it have a lithe, subtly striped trunk and beautiful flowers, but it produces sweet dark red fruits that are a bit like the rose family's answer to the blueberry — similar in size and appearance, but with more apple and almond in the flavor. They range from tasty to fantastic. Some suggest that *Amelanchier laevis* (see below) tastes even better, but there is also widespread acknowledgment that the genus is a taxonomic muddle. My advice is to try any wild species that range in your region and find what works best in terms of flavor and habitat tolerances.

Smooth serviceberry (*Amelanchier laevis*) has some of the tastiest fruit in this diverse genus. An understory tree, smooth serviceberry appreciates soil moisture and is found in a variety of habitats from pond and swamp edges up to high elevation openings in the Appalachians.

In addition to being one of the tastiest of the serviceberries, smooth serviceberry also shines in the ornamental category, with relatively large white flowers in early spring, and elegant trunks that feature subtle vertical striping. Newly emerging foliage has a coppery tone and lacks hairs on the leaves.

The fruits are sweet, soft, and delicious, with a hint of apple and almond flavor. A variety of serviceberry species are used by Indigenous peoples to make pemmican, a mix of berries, dried meat, and animal fat

Smooth Serviceberry (*Amelanchier laevis*).

that keeps well. With the resurgence of gourmet and novelty jerkies and the like, can one made with the traditional native ingredients of serviceberries or chokecherries be far behind?

Similar to the above two species is **Canadian serviceberry** (*Amelanchier canadensis*). This species is most common in coastal and sandy areas. It is frequently used in the landscape trade due to its relatively compact, multi-stemmed form. In addition to the absolutely delicious fruit (in Maine they call this and related species "sugar pears"), this is a slender-trunked, elegant shrub with pure white flowers early in spring.

Running juneberry (*Amelanchier stolonifera*) is more of a shrub than the species described above. Sometimes it is quite diminutive; the first time I saw this plant it was running colonially along a crack in a massive sheet of bedrock on the side of Jenny Jump Mountain. Its low stems rose a foot or so above the shallow crevice tracked by its stolons and roots. It has small, sweet fruit similar to other serviceberries, with overtones of apple and almond. I think of it as a glade and barrens specialist, but others note that it can also be found in open fields, fencerows, and other similar habitats.

Mountain serviceberry (*Amelanchier bartramiana*) is a relatively short, shrubby serviceberry that thrives along moist edges with a fair amount of sun. It is native to the coldest parts of our region. It differs from some of the other serviceberry species described here in having somewhat oblong, pear-shaped fruits and flowers produced in small clusters. Some describe the fruit as sweet and dry, others as sweet and juicy. Like many in the genus *Amelanchier*, some variability may be present. Mountain serviceberry forms loose colonies by layering stems.[60]

Amphicarpaea bracteata • Hog Peanut

Quick leguminous groundcover with tasty little "beans."

This versatile legume is a scrambling annual vine that thrives in moderately disturbed shady habitats. Small purplish flowers form pods resembling small pea pods, containing a small, edible legume. These usually dangle from some taller herb or woody plant stem that hog peanut has (harmlessly) twined itself around.

Additional, belowground flowers form larger seeds that are a more choice edible. Due to the small size of these seeds and the necessity to search and dig for them, it can be challenging to harvest significant quantities. Santee Dakota women in what is now Minnesota used a three-foot long digging stick, with a bowl-shaped end, made of ironwood, to dig for the seeds. However, the primary method in which these legumes were sought was to raid the caches of the "bean mouse" — the vole *Microtus pensylvanicus* and other rodents — often leaving some food item in trade.[61]

Given that hog peanut is an annual and a nitrogen fixer, this species makes sense as a restoration species for shaded areas that need stabilization, weed suppression, and soil improvement, as perennial plant assemblages develop.

Anaphalis margaritacea • Pearly Everlasting

A sage white smudge brings ease.

Covered with a fine pearly white down, with largely white, clustered flowers, this plant does indeed have a luminous sheen. It is a medium-height herb of open, dry, often lightly disturbed habitats. Presumably its white hairs help retain a moist micro-environment around its photosynthetic tissues, relieving desiccation pressure. Its traditional herbal uses range from the mundane (as a soothing expectorant) to the physically powerful (to revive those suffering from paralysis) to the ceremonial (some northern Indigenous cultures use pearly everlasting as a smudge in a capacity similar to the incense of white sage, often in combination with tobacco). Other herbal uses include as an internal or external astringent (diarrhea, burns), as a cooling herb for fever, and to relieve headaches.[62]

The flowers are said to assuage thirst on long hikes. They last for many months in dried bouquets, hence the name "everlasting." Swedish botanist Peter Kalm reported it growing "in astonishing quantities upon all uncultivated fields, glades, hills, and the like" in what is now Pennsylvania in 1748,[63] probably as a result of widespread burning[64] by Lenni Lenape of the region.

Pearly everlasting (*Anaphalis margaritacea*).

Angelica atropurpurea • Purplestem Angelica

Gigantic herb with powerful root.

Purplestem angelica is a giant of a flower, with blooms in a compound umbel that radiates in all directions like an exploding firework. It is found in sunny riverine wetlands with calcareous geology.

Its young shoots and leafstalks can be cooked as a celery-like shoot vegetable, or candied. Some members of the Apiaceae can cause photodermatitis due to furocoumarins they contain. I experienced this with angelica: after candying some stems in honey one morning, that evening I felt like I had sunburn or windburn on my face, despite not having been overly exposed to the elements.

The medicinal root, after drying, is said to resemble its European cousin, *Angelica archangelica*, in medicinal application. Herbalist Matthew Wood utilizes the native species. He describes angelica as "pungent, bitter, sweet, oily, warm, and stimulating" to the digestive, circulatory, and lymphatic systems.[65]

Apios americana • Groundnut

Combining the best of potatoes and beans, with tubers strung on underground cables.

Groundnut is a twining herbaceous vine arising from underground tubers connected along stringlike rhizomes like highly irregular holiday lights. It bears elaborate, salmon-and-ochre blooms in showy clusters that form elongate pods like narrow green beans, complete with small leguminous seeds within.

The edible portions of the plant are its fleshy tubers, which resemble potatoes in form but have a more bean-like or nutty flavor. These are produced in abundance from an early age, are pleasing to the palate, and have high potential as a food staple. They taste especially good fried in oil in a skillet, where they form a crispy, flaky skin, something like a gourmet native Tater Tot. They can also be ground for flour to make bread. Groundnut tubers contain 13–17% protein, about three times that of potatoes and other domesticated root vegetables.[66]

The plant grows rapidly from seed and even more quickly from tubers, and can be readily established in areas with adequate moisture. In deep, moist soils, rhizomes spread deeply and at some distance from surface stems.

Unlike some vines, groundnut is relatively modest in its aboveground growth and coexists well with diverse plant associates. Its most frequent wild trellises are shrub species such as winterberry holly, spicebush, and the like, but it will scramble and climb at will. Elderberry makes a logical associate when creating edible habitats, as do upright forbs ranging from common milkweed to swamp rose mallow. Groundnut can fix nitrogen, potentially benefitting soil fertility.

Early accounts suggest that groundnut was cultivated by Indigenous people in the early colonial era. Lescarbot described groundnut planted in northern New England in his 1609 treatise *Nova Francia*.[67] The vast majority of groundnuts found north of Connecticut are sterile triploids, reproducing only vegetatively.[68] One can conjecture that these plants were in fact selected and spread by Indigenous peoples up and down the New England coast. Along the Androscoggin River in Maine I recall finding an exceptional growth of groundnut, stretching hundreds of feet, acting as a vigorous groundcover, and possessing very large tubers.

Apocynum androsaemifolium • Spreading Dogbane
Apocynum cannabinum • Indian Hemp

Cordage plants of the field.

Indian hemp and dogbane, the two common names for our native *Apocynum* species, point to important attributes of the characters of these two related species. The name Indian hemp tells us about one of the strongest, most workable wild cordage materials on the continent, used for everything from bowstrings to clothing. The name dogbane, on the other hand, reminds us of the toxicity of the plants in this genus.

Many poisonous plants have a dual nature, as toxic agent and healer. Herbalist Matthew Wood calls spreading dogbane (*A. androsaemifolium*) "Werewolf Root" and uses it for people who "need to make a ninety degree turn or they will die — spiritually or physically." Note that he uses the plant only homeopathically, not in material doses. If you're curious, I suggest you read more in his excellent and highly narrative tome *The Book of Herbal Wisdom*.

Both dogbane species contain toxic cardioactive glycosides. That said, there are no known cases of fatalities due to dogbane poisoning, and historical medicinal uses for both species are well-attested. *A. cannabinum* was listed in the U.S. National Formulary as a cardiac stimulant, and *A. androsaemifolium* was described as a cathartic, diuretic, diaphoretic, emetic, and expectorant.[69]

Indian hemp is an aggressive spreader and can create large colonies in situations where competition is limited. It is often an early colonizer of fields left to fallow or abandoned from farming.

Aralia nudicaulis • Wild Sarsaparilla

Colonial herb with potent roots and fruit.

One of the charms of old time sodas and candies was that they were fashioned from all manner of phytochemically potent roots, saps, barks, and blooms. Hence slippery elm throat lozenges, birch beer, black cherry cough drops, and root beer. Many native and imported medicinal herbs such as sassafras, spikenard, licorice, ginger, and dandelion contributed to root beer's rich flavor and its tonic powers.

Wild sarsaparilla was one of these root beer herbs, used as a substitute for the imported sarsaparilla. Even before that, it was an element of Indigenous medicine, said to give strength to the weak, weary, and old. In that matter, its usage is quite similar to its relative ginseng, which it also resembles in vegetative form to a striking degree, though with pinnate leaves instead of palmate. Other uses include as curative for infections, and to reduce swelling.[70]

Its fruits are an overlooked item, very sweet, with a blackberry-ish or currant-like flavor, and with the soapy finish that is characteristic of the genus *Aralia*. Some sources mention making the fruit into jelly or wine.[71] Like its relative *Aralia racemosa* (see below) its young shoots can be eaten as a cooked vegetable, and its roots are edible as well.

Aralia racemosa • Spikenard

Giant forest herb forests yields shoot, fruit, and root.

Once in a great while I chance upon spikenard in the wild, and it always seems to be an indicator for a special place. I doubt I've ever seen it any place where the remainder of the plant community was not in some way interesting and diverse. Hanging from a shaded dolomite cliff. On the banks of a cool mountain stream. At the base of north-facing shale outcroppings. In a

Spikenard (*Aralia racemosa*).

diabase ridgetop forest. The species more or less exemplifies the concept of conservatism expressed in Floristic Quality Assessment (FQA) and the idea that communities of specialist plants cluster in special habitats (see the Appendix for more on FQA; hope I got you intrigued).

Not to suggest that spikenard is in any way fragile or persnickety. On the contrary, this is probably the largest perennial forest herb in our flora, sometimes taller than a human being, with an architecture and massive foliage to match. Spikenard's thick, ropy rhizomes radiate octopus-like from the crown. Inside, they are sweet, balsamic, sticky, with a bit of the acrid finish typical of *Araliaceae*. These are the parts used in herbal medicine.

In the 1920s and early 1930s, botanist and ethnographer Huron Smith studied medicinal plant uses among six Indigenous cultures in Wisconsin. He was able to complete four ethnobotanical studies before his tragic death in an automobile accident in 1933: treatises on the Menominee, Meswaki, Ojibwe, and Potawatomi. His work on the Oneida was compiled posthumously in 1998. Only field notes and herbarium specimens exist for his work among the Ho-Chunk (Winnebago) until their recent compilation by Kelly Kindscher and Dana Hurlburt in 1998.

Smith was exceptional in that he combined exemplary botanical skills with ethnographic approaches, and was an astute student of herbal medicine. He was also a respectful participant-observer of Indigenous culture and was a quick study in languages, allowing for a deeper rapport with his subjects.

Each culture that Smith studied with utilized spikenard in different ways, with some overlapping themes. Combined, the uses documented by Huron Smith offer a broad picture of the powers of the herb.

Areas of overlap include two cultures utilizing spikenard as stomach medicine, the Ho-Chunk and Menominee. Two cultures had uses pertaining to blood — as a purifier during pregnancy among the Ojibwe, or for blood poisoning among the Menominee. The theme of pregnancy was also present in Meskwaki practice, "where mothers [or infants] have it 'showered' or sprayed from the mouth upon their heads, when they are giving birth [or born]." Three cultures used the root externally, the Ho-Chunk for "boils, sores, and carbuncles," the Menominee as a poultice for sores, and the Forest Potawatomi as "a pulp to be used as a hot poultice on inflammations." Among the Ojibwe, a hot infusion of the pounded root was drunk for coughs.[72]

Spikenard is an herb with well-attested traditional uses, held in high regard by some contemporary herbalists, but also obscure to some degree. It has never reached the level of contemporary popularity and regard as other native forest herbs that share its habitat have, such as ginseng, goldenseal, and black cohosh. With the latter becoming faddish just in the past few decades, and facing potential depletion in its natural habitats due to overharvesting, this is probably a blessing for spikenard. However, it could be just a matter of time before such a powerful, broadly useful medicinal herb is re-discovered by industry and marketed to a public hungry for natural solutions to health issues.

Spikenard also has edible fruits, sweet, spicy, and with a soapy finish. Both these and the rhizomes have food potential but probably require some innovation in terms of preparation to render them choice to the contemporary palate.

The most choice edible part of spikenard is its shoots. The herb arises from its winter dormancy like a vegetal bear paw reaching for the canopy. These ursine shoots, with leaves still tightly furled, make an excellent, meaty spring vegetable. Each plant can tolerate a moderate amount of shoot harvest each year, in a similar fashion to a stand of asparagus, or giant Solomon's seal.

Arctostaphylos uva-ursi • Bearberry

Groundcover for sterile soils, smoking mixes.

Bearberry is a small, glossy-leaved groundcover, an evergreen subshrub that has an affinity for dry, sandy soils as well as thin soils over acidic bedrock. It is a circumboreal species, found in cold climates across North America, Asia, and Europe. It extends south into the New Jersey Pine Barrens near the coast, and further south in the mountains. It is a slow spreader, but capable of covering large areas of infertile soils over time. The fruit is edible (best cooked), and the foliage is used for tea and is added to Indigenous smoking mixtures. Among the Arikara, "The dried leaves of this plant were an important part of the smoking mix (*kinnikinnick*), with it and the inner bark of red osier dogwood, (*Cornus sericea*), being added to the leaves of the special tobacco (*Nicotiana quadrivalvis*) that the Arikara cultivated for its use," reports early ethnobotanist Melvin Gilmore.[73]

Aronia melanocarpa • Black Chokeberry

Antioxidant superfruit.

This shrub has an interesting plasticity of form and range of habitat. It is found both as a low shrub in xeric habitats, and as a taller shrub in wetlands. It seems not to be particularly competitive in mesic conditions, and is intolerant of shade.

Aronia fruits are astringent and thick-skinned but make excellent juices, syrups, and fruit leathers — they're just not choice for fresh eating. About the size of a blueberry (and as variable), these matte black fruits grow in dense clusters sometimes heavy with fruit.

Fans of Aronia fruit face a double challenge in its marketing. Its common name, chokeberry, does not give it instant appeal. And its highly astringent skin makes that name somewhat apt. The solution has been to re-brand it as "Aronia" fruit and popularize its superfruit status. Both appellations are apt, and Aronia is off the charts in terms of antioxidant content, due to the anthocyanins in the fruit's skin. These help protect against a variety of degenerative diseases.[74] The fruits also contain high levels of fiber and vitamin C. The fruit is useful for maintaining urinary tract health[75] and as an anti-inflammatory. Polyphenols in the fruit support cardiovascular health.[76]

The flavor of the raw fruit can be variable. Because of its intense tannins, Aronia is often consumed as a juice, syrup, or jelly. Though it may appear ripe by mid-summer, it is best harvested later in the season, when it is sweeter and less astringent.

The closely related **red chokeberry** (*Aronia arbutifolia*) is also edible and features high anthocyanin levels.

Arundinaria gigantea • River Cane
Arundinaria tecta • Switch Cane
Native bamboo.

Tall, thick stands of river cane once lined the waterways of the Southeast, ranging from southern Virginia to Florida. These native bamboos wove through the heart of Cherokee territory on myriad streams and were in turn utilized by Cherokee women (and those of other tribes) for weaving a great variety of baskets, mats, and ceremonial rugs.

In her book *Weaving New Worlds: Southeastern Cherokee Women and Their Basketry,* ethnographer Sarah H. Hill catalogs the numerous uses of this utilitarian plant. It was used in the manufacture of house walls, hair ornaments, game sticks, musical instruments, toys, weapons, tools, and beds. As for edible uses, "[W]arriors purified themselves with cane and root tea" and Cherokee "made flour from cane," Hill writes. The young shoots are also edible, like the Asian bamboo species that are their close relatives.

River cane has greatly diminished in abundance. The once extensive stands of river cane can be ascribed in part to Indigenous management.[77] River cane is abetted by fire, and also by flooding. These disturbances clear competing vegetation as well as accumulated leaf litter. In the absence of these disturbances, stands reach maturity and then begin to decline after they are a decade old.

River cane is sensitive to grazing, and was much affected when cattle pasturing was introduced by European settlers. Canebrakes on fertile alluvial soil were identified as ideal habitats to clear for crop farming. Together with fire suppression, these activities have led to a substantial decline in river cane habitat. While river cane rhizomes can be long-lived, awaiting release from a canopy-clearing disturbance, recruitment from seed is slow and uncommon.[78]

The unique habitat created by cane supports a number of warblers, including the possibly extinct Bachman's warbler, which depends on canebrakes for nesting habitat and whose disappearance is subsequent to the decline of river cane habitats following the replacement of Indigenous cultural landscapes by European-derived agricultural ones.

Up to three species of cane are currently recognized, though taxonomy is in flux and it is acknowledged that growing conditions can affect the size and other characters of *Arundinaria* species. *Arundinaria gigantea* is the primary species discussed here, found in the Piedmont and mountains from Delaware and Virginia south. Its hollow stems can be 30 feet tall and up to three inches thick. Switch cane (*Arundinaria tecta*) resides in coastal plain wetlands. Hill cane (*A. appalachica*), a species of the southern Appalachian uplands, is outside the range of this book. It is quite diminutive in size, usually one and a half to three feet tall. It was only recently recognized as its own species.

Asarum canadense • Wild Ginger

Shining hearts arising from a spicy sweet rhizome.

With heart-shaped foliage horizontal to the ground, and a light iridescence, this species forms a beautiful, low groundcover, creeping on slender rhizomes between rocks and tree trunks. Its improbable-looking, highly geometric flowers are stiff barrels, mainly corolla, patterned with purplish brown. They lay directly atop the soil surface, concealed except to those who part the leaves to search beneath.

Used as a flavoring and tea, its rhizomes are known in ethnobotanical literature as well as among foragers. Its fragrance is one of my favorite native plant aromas, up there with the flowers of stoneroot. Wild ginger was used extensively by Indigenous people for medicinal uses.[79] Increasing caution has limited its contemporary use due

Wild ginger (*Asarum canadense*).

to the presence of aristolochic acid, a carcinogen. That said, no study has assessed the possible carcinogenicity of wild ginger as it might actually be used in a human diet or medicine. Because aristolochic acid is only slightly water-soluble, I continue to use this plant in a tea.[80] Make an appropriate decision for yourself based on current research.

Regardless of edible uses, this makes an excellent and surprisingly versatile groundcover in shade, beneath larger perennials and woody plants.

Asclepias syriaca • Common Milkweed

Break bread with a monarch.

This tall herb is adapted to disturbed habitats. It is especially common in formerly plowed fields, as plowing breaks its spreading rhizomes and disperses common milkweed across the field. It is the host plant for monarch butterfly caterpillars, which can only consume plants in the genus *Asclepias* (i.e., monarchs are a specialist on milkweeds). With large, honey-scented spherical clusters of flowers, it is also a showy wildflower and a terrific pollinator plant.

There are two camps regarding the edibility of common milkweed. The status quo suggests this is a toxic plant, imbued with the cardiac glycosides common to much of the genus. Powerful toxins derived from some species in this genus have been used as arrow poisons by Indigenous peoples in South America and Africa.

Therefore, some foragers treat milkweed as toxic and bitter. They recommend cooking the hell out of milkweed, boiling in multiple changes of water before consumption.

Others dissent. Sam Thayer, who gives milkweed (and this issue) an excellent treatment in his book *Forager's Harvest*, suggests that reports of milkweed toxicity are exaggerated and pre-boiling in changes of water is unnecessary. Meanwhile, ethnobotanist Kelly Kindscher recommends cooking for four minutes and changing the water at least once prior to consumption. I've never detected bitterness or toxicity in my cooked milkweed. I usually boil it briefly, discard the water, then transfer the shoots to a cast iron skillet to braise to a finish. Read around, try it, and decide for yourself.

Based on his fieldwork among the Arikara in the early 1900s, Melvin Gilmore records: "This is a useful food plant. The young sprouts, the tender young leaves and tips, and the bud clusters and young seed pods were stewed either alone or with green or dried corn or with buffalo meat." In his 1926 article *Some Interesting Indian Foods*, Gilmore wrote: "Common milkweed was also important to making traditional corn bread. Ripe corn was pounded to meal with a pestle in a mortar. The meal was mixed with water and seasoned with salt. The fireplace meantime had been well heated by maintaining a hot fire for a sufficient time. Then the coals and ashes were drawn out, a layer of fresh green leaves of the common big milkweed (*Asclepias syriaca*) was laid in the bottom. The corn meal batter was poured upon this layer of fresh, clean leaves. Then another layer of milkweed leaves was laid over the batter. Then a layer of corn husks was laid over the leaves, then a thick layer of ashes over the corn husks. Last of all a good bed of live coals was laid over the ashes, and the bread was left to bake." The milkweed leaves added moisture and likely flavor to the baking bread.[81]

If you've done your research and would like to try it, here is my take. Milkweed is a superb vegetable: flavorful, tender, abundant. It is used like asparagus (as shoots), broccoli (unopened seedbuds), and okra (unripened seedpods). It may be better than any of the above ... and it's perennial, so ground does not need to be freshly disturbed, nor new seedlings started every year, to enjoy it as a vegetable.

As anyone who has seen a cut hayfield with a late bloom of milkweed knows, milkweed is quite capable of regenerating from serious trimming, so a modest harvest can be considered to be completely sustainable. An ideal permaculture crop then; and a forager's delight as well.

Milkweed is neither scarce nor slow-growing, but it is part of a wild

Common milkweed (*Asclepias syriaca*).

commons we share with other animals. What if we planted milkweed in perennial beds near, but not in, our vegetable gardens or annual farm rows? We could give freely to monarchs and other pollinators, and take freely for ourselves as well. This is the point where farming dissolves into ecological restoration, and this is where the native plant movement can go: supporting the needs of all animals, human and otherwise, in a new, abundant commons.

Asclepias tuberosa • Butterfly Milkweed

Blazing orange flower likes blazes, cures fire in the lungs.

Nowadays butterfly milkweed is mainly noted for its glorious orange color and extended bloom time. Like other milkweeds, it is a pollinator magnet, drawing swallowtails, fritillaries, and a host of native bees.

Butterfly milkweed has a deep taproot with dormant buds well below the soil surface. It responds to fire with a blazing show of its own, sprouting prolifically from these buds after blazes. This is a very long-lived perennial but relatively short in stature, so it relies on poor soils and fire to keep its competition down, or it may fade from a site.

In a former life, butterfly milkweed was known as pleurisy root and was listed in the *U.S. Pharmacopeia and National Formulary.* Pleurisy refers to an inflammation of the pleura, a layer of tissue surrounding the lungs, and the dried root of this milkweed is used for acute infections of the lower respiratory system, with uses for bronchitis, pneumonia, and coughs. In addition to these internal uses,

several Indigenous peoples utilized the roots for wounds, sores, cuts, and bruises.[82]

Contemporary inquiry into pleurisy root suggests activity from cardenolide glycosides and flavonoids which may contribute to its traditional uses for pneumonia and as an analgesic, anti-inflammatory, diaphoretic, and digestive agent.[83] Toxicity may be present as well due to the presence of glycosides.

Asimina triloba • Pawpaw

Lush and sheltering, this fruit-bearer links us to far-off, tropical climes.

Pawpaw is a distinctive small tree with a tropical appearance and fruits. It has large, lush leaves, and its edible fruits somewhat resemble mangoes in shape. It grows in a range of habitats but is most closely associated with floodplains and other moist, fertile sites, including mesic forests and coastal swamps. Pawpaw grows colonially and can persist as a low shrub in the shade for long periods waiting for a canopy gap. Given adequate sunlight, it grows larger in size and commences more prolific fruiting. If growing this species for fruit, a relatively sunny location is advised. Note however that first-year seedlings are intolerant of full sun exposure and are generally grown under shade in nursery practice.

It is hypothesized that the primary dispersers for pawpaw seeds were now-extinct Ice Age megafauna. Only these large beasts could ingest a fruit whole, defecating out the large seeds in a new location. Other than flooding and the occasional transport of a few seeds by an existing large mammal, we humans may be the primary seed dispersers for this fruit until such time as North American megafauna return.

Pawpaws have become quite popular in recent years. If this is your first time hearing about them, the fruits have a tropical flavor, and can be quite sweet and delicious. They are somewhat variable in flavor, texture, and color. They do not store well, and do not taste good overripe. Personally, I'm happy to eat a couple of pawpaws every season and leave the rest for wildlife but many other people feel differently. I once tasted a truly delicious pie made with pawpaw; other prepared foods have been less appealing. However, there are a good number of pawpaw obsessives out there constantly innovating with varietal selections as well as food preparation methods, so stay tuned.

Pawpaw produces defensive chemicals unique to the custard apple family, and few wildlife species consume its foliage. It is, however, the sole host plant for the zebra swallowtail butterfly.[84]

Betula: The Birches

Tea trees, fire-starters.

The small, wind-borne seeds of the birch family give us a clue to their ecological role as land healers. Easily blown in from parent trees at some distance, birch seeds recruit best on exposed soils such as those created by logging, fire, windstorms, and other large-scale disturbances.

The birches discussed below inhabit a variety of habitats from mesic mid-Atlantic woods to the boreal forests of the far north. As land healers, they are essential tools for the restorationist. As medicine plants, they offer a balm for our human bodies via methyl salicylate, the compound from which aspirin is derived. Their contributions as food plants and craft materials cannot be understated either, with evidence of the use of birch tar by Neanderthals (in Europe) as much as 80,000 years ago![85]

With their thin, highly flammable bark, individual birches are quite susceptible to wildfire. However, they are also poised to take advantage of post-fire conditions by germinating readily in bared soils.

Yellow Birch (*Betula alleghaniensis*): Yellow birch is beautiful when young, dramatic when old. Youthful yellow birches have tight, metallic bark and thin, graceful twigs. When older, the bark peels in strips of bronze, yellow, white, gold, and silver.

Young trees often recruit on leaf-litter-free surfaces, including the stumps, nurse logs, and tip-up mounds of trees that have gone before them, releasing extra light to the forest floor with their demise. Birches that germinate on fallen and rotting wood first root into their immediate substrate, but also send roots down towards the

Yellow birch (*Betula alleghaniensis*).

the tree generally grows on mesic to wet sites, favoring northern or mountainous locations. Yellow birch's foliage hosts abundant insect life making it an excellent species to support songbirds.[86]

Like its relative, black birch, yellow birch's sap (and therefore cambium) contains wintergreen (methyl salicylate) and its twigs make a delicious tea as well as a remedy for sore muscles, spasms, and arthritis, and are used as well for digestive issues and as an antiseptic.[87] Its sap can be tapped for syrup or used to make hand-crafted, all-natural birch beer. That would be a great premium product for a rural, native-plant-based farmstead! The sap drunk fresh is faintly sweet and glows with vitality, like an ent-draught from Middle Earth.

The peeling bark of yellow birch makes an exceptional fire starter, easily ignited even in wet conditions. It shreds easily to create tinder for the ancestral skills practitioner or the camper trying to start a warming fire on a rainy day.

Black Birch (*Betula lenta*): When I pass a black birch growing along an edge with its branches hanging low, I'll generally snap off a twig about as wide as a pencil, strip the leaves, and hand out a section to anyone else walking with me. Right away the scent of sweet wintergreen wafts up from the aromatic sap.

If we're hiking, the twig becomes a chew stick. Once masticated, the twig ends

ground. When the stump or nurse log the birch germinated on rots away, the birch's roots remain, octopus-like, supporting a tree that seemingly germinated in mid-air.

Despite its conspicuous roots, yellow birch's overall root system is shallow, and fray into a brush of sorts, one flavored with natural wintergreen and sweetened with xylitol, a sugar alcohol produced by black birch. The result is not too different from a toothbrush and toothpaste, but far more versatile in terms of gum stimulation and without all the plastic and industrial chemicals. For most people, this will remain a curiosity, partially because black birch largely loses its aromatic properties on drying, so twigs can't be hoarded for long as biodegradable toothbrushes; you need to procure a fresh supply from a living tree with some regularity.

Black birch's aromatic oils also lend flavor to birch sap and syrup, the former a refreshing spring drink straight from tapped trees, the latter distilled in the manner of maple syrup. The wintergreen essence extracted from the tree is used as a topical analgesic, as well as a flavoring for chewing gum and the like.

Birch also makes a delicious tea. Twigs are infused in hot water (not boiled) until the water takes on a pink hue. The easiest and least harmful way to prepare it is to cut twigs from a tree and use clippers to render them into small sections to infuse with. The flavor intensity of birch is highest at the times of year when the sap is flowing. The twig sections must be used fresh, as the essential oils dissipate fairly rapidly.

Black birch is a gap colonizer in forests, often recruiting when mature trees die, are wind-thrown, or after logging occurs.

While not a fire-tolerant species, the exposed mineral soils after burns make an ideal ground for the germination of birch's abundant, wind-borne seeds. Though the seeds are tiny, black birch seedlings are capable of rapid growth and can be several feet tall after two seasons.

A variety of birds consume the catkins, seeds, and buds of black birch, parts that are often available in the winter at a time when other food can be scant.

Paper Birch (*Betula papyrifera*): One of the iconic trees of the far north, this white-barked birch grows from Newfoundland westward along the northern boundary of tree growth, extending southward into the northern states of the continental United States. Paper birch can form pure stands on disturbed soils, resulting in visually stunning groves of slender ivory birches.

The bark of paper birch is both flammable and also waterproof. It has a long history of use for containers and canoes. The sap is used for syrup, fermented beverages, and in medicinal tonics.[88]

Paper birch, like the other birches above, is a relatively short-lived tree. Its primary restoration use is for soil stabilization and revegetation of highly disturbed sites, including mine spoils and other areas where mineral soils are exposed. Over time, paper birch stands may be replaced by spruce and fir on thin soils, or by later successional northern hardwoods.[89]

Cakile edentula • Sea Rocket

Seashore spiciness.

Sea rocket is a maritime mustard that occupies the high beach area, seaward of dunes, and is found sporadically on back dunes as well. Some seed is dispersed in floating pods and capable of travelling long distances,[90] while other seed is sent landward via wind for contingency populations.

The leaves are delicious raw: succulent, crisp, slightly sweet, with a horseradish bite to the finish. The young seedpods are somewhat pea-shaped and reminiscent of peas in being juicy and sweet — with the same spicy finish as the leaves.

Caltha palustris • Marsh Marigold

Boiled spring green.

This is a luminous early spring wildflower with big yellow blooms and vaguely heart-shaped leaves.

Those leaves are used as a spring green, usually boiled. This preparation is necessary due to the presence of terpenes that are gastrointestinal irritants, but are rendered safe by heating. They are safest eaten before flowering, and are to be avoided after the blooming season due to the accumulation of toxins.[91]

I usually see marsh marigold growing in forested wetlands and along shady streams, but farther north it becomes more prevalent in sunny habitats.[92]

Note that marsh marigold can be mistaken for the extremely invasive groundcover lesser celandine (*Ficaria verna*).

Cardamine bulbosa • Spring Cress

Sweet little mustard.

I consider this seldom remarked-on flower to be both a visual and culinary treat. Its cheerful bright white blooms enliven forested swamps and stream edges in early spring. A diminutive plant, it is often found growing right on mossy, moist boulders. Its foliage tastes a lot like wasabi — sweet and pungent. Some suggest using its roots as a native horseradish but I've never sampled this part of the plant. In full sun, it can be fairly productive. Usually I see it in shadier nooks where it can avoid competition from larger species.

Cardamine concatenata • Cutleaf Toothwort

Wild wasabi of spring.

For about six weeks in spring, this low herb sprouts a few roughly cannabis-shaped leaves, goes to flower, catapults seeds from thin siliques, and goes dormant until the following April. Sharing a lifestyle with other spring ephemerals like spring beauty and trout lily, this plant is a diminutive mustard called cutleaf toothwort. Its foliage (and roots) bear the flavor of horseradish and wasabi — delicious if you like some sweet heat with your sushi, sandwiches, or sauces.

Cutleaf toothwort can grow in large colonies — a former argillite quarry downriver from me is covered in the stuff — but mostly this herb is interspersed in small groups throughout suitable habitat. Because of this, and its modest stature (usually under a foot tall), it is unlikely to feed the masses any time soon. But adding it to your restoration brings not only spice to life but also supports early spring pollinating bees who appreciate toothwort's simple, four-petalled pinkish-white blooms.

Also edible and fulfilling similar ecological roles are *Cardamine diphylla*, *C. maxima*, and *C. angustata*.

Carya: **The Hickories**

Eccentric and nuts.

Shagbark Hickory (*Carya ovata*): Imbued with visual charisma, the shagbark hickory is a tree easily recognized once you've been introduced. Most conspicuous is its namesake bark, which cloaks the tree in great ragged strips with an eccentricity that makes each mature tree stand out. The peeling bark, elegant in its way, clads a stout stem of some of the hardest wood on the continent. Following this wild and stout aesthetic, its branches radiate and fork like arboreal lightning.

Shagbark hickory nuts will be familiar in flavor to anyone who knows pecans, a close enough relative in the same genus. To my taste, shagbark hickory nuts have a superior and more concentrated flavor. Traditional usage of these nuts extends beyond just raw snacks, to high quality oils

as well as flours. The calorie, lipid, and protein content of hickory nuts is far superior to acorns, and on a par with or better than black walnut, butternut, and American hazelnut .[93]

Shagbark hickory (*Carya ovata*).

In addition to nuts, shagbark offers a hard, dense, aromatic wood useful for tools, smoking food, and some of the highest BTU firewood around. It also produces a delicious, aromatic, almost smoky syrup produced by simmering chips of roasted bark covered in water and adding a sweetener (i.e., not by tapping for sap like a maple). Modest amounts of this bark can be sustainably harvested from living individuals.

Shagbark can be slow growing at first; much of its energy is concentrated on taproot development. It is a species that puts effort into a quality foundation and intends to stick around for the long run. Shagbarks are long-lived and fairly shade tolerant. They are somewhat less fire-tolerant than oaks, but the shagginess of their bark can insulate the trunk from low intensity fire damage. Larger trees (over ten inches diameter at breast height) are less vulnerable to fire than seedlings and saplings, which are frequently top-killed. Even though fire may cause stem mortality, shagbarks of all ages will often send up fresh sprouts from their root systems.[94]

Hickories are a good example of a species documented as encouraged near historical Indigenous settlements. For example, Cherokee women maintained clearings to encourage hickory and other nut- and fruit-bearing trees. Hickories are most productive in open habitats, yielding eight times more nuts than in closed canopy forests.[95] When I forage for hickories, I often look for trees on the edge of country roads, yards, or grown in open settings in parks.

Shellbark hickory (*Carya laciniosa*) is a similar species with a somewhat more limited native range. Nuts are even larger than those of shagbark hickory, and their flavor is excellent. It has an affinity for moist bottomland forest habitats.

Mockernut and pignut hickories (*C. tomentosa* and *C. glabra*) vary in palatability as raw nuts. I usually find them decent to eat, but not nearly as good as shagbark. I don't have experience with **sand hickory** (*Carya pallida*) but coastal plain dwellers take note (and report back, please). **Bitternut hickory** (*C. cordiformis*) has bitter tannins that make the nuts unpleasant for raw eating. However, bitternuts contain 60–70% oil, and the bitter tannins are water-soluble, so the oil is perfectly free of them.[96]

Castanea dentata • American Chestnut

The once and future forest food.

I finally saw a full-size American chestnut tree in the woods last year. Straight as a tuliptree, clear of branches for 30 feet, it stood among mixed oaks along a woodland trail. To be clear, I've seen hundreds of American chestnuts, but they've always been multi-stemmed suckers coming off of ancient root systems whose main stems were stricken in the early 1900s by the chestnut blight. To just stumble on a healthy, canopy-high tree along a trail I'd walked several times before ... it was like seeing a ghost.

I've never tasted an American chestnut, but if the flavor of imported chestnuts, or the reverence that this tree was given before being laid low across its range by the imported blight are any indication, this must be a delicious, sweet, sustaining nut. It is a prolific tree and its nuts are reportedly easy to make into any number of foods, from chestnut flour to the proverbial roasted chestnut. Chestnut bread and stored nuts were an important winter staple for the Cherokee and many other tribes. Chestnut was a dominant tree especially in the Appalachians, where mature individuals could reach a height of over 120 feet and a girth of seven feet in diameter, "blackening the ground" with nuts in autumn.[97]

We don't fully understand the fire ecology of American chestnuts. They can certainly resprout vigorously after the tops are killed. However, they may be less tolerant of fire than their frequent companions, the oaks.[98] William Doolittle, in *Cultivated Landscapes of Native North America*, describes an intriguing but unverified account from the 1800s, where "remnant groups [of Indigenous peoples] in Massachusetts [went] to some length to protect chestnut groves from annual fires that presumably were intentionally set."[99]

American chestnut (*Castanea dentata*) hybrid.

Researchers with the American Chestnut Foundation are working diligently to create disease-resistant American chestnuts, and we are getting closer to the dream of reintroducing this abundant, delicious, and ecologically foundational tree back to its native haunts.

Castanea pumila •
Allegheny Chinquapin

Little chestnut shrub.

Chinquapin is a little sibling of the American chestnut, bearing smaller nuts. Some say it is delicious, others that the fruits are frustratingly small. Though it is susceptible to chestnut blight, as a multi-stemmed shrub, it can sucker freely and still reach reproductive maturity on some stems before they succumb to blight and new sprouts appear. Reports vary on its susceptibility to the blight, and it may be that there is some regional variation in this regard.

Chinquapin grows along the mid-Atlantic coastal plain and into the Appalachian Mountains, native from New Jersey southward. It thrives in dry, sandy, or rocky soils, in plentiful sunlight. According to the U.S. Forest Service, "Chinquapin forms extensive clones where it has been burned over annually or at short intervals." It benefits from fire that thins overstory competition; though if chinquapin's aerial stems are killed, it resprouts and is rejuvenated after burns.[100]

Caulophyllum thalictroides •
Blue Cohosh

Birthing medicine.

Blue cohosh is a graceful herb with a horizontal compound leaf borne atop a slender stem, a form typical of a number of woodland herbs. It looks just a little bit vulnerable, but this architecture has proven itself over the millennia and large colonies of this herb thrive in moist, rich woods.

Traditionally, blue cohosh root is used to ease labor and normalize menstruation. Indigenous groups such as the Cherokee, Chippewa, and Ojibwa use the plant to reduce cramps.[101] Herbalist and grower Richo Cech writes of its use during childbirth: it "stimulates the pituitary gland to signal

Blue cohosh (*Caulophyllum thalictroides*).

increased production of ... oxytocin, a hormone that stimulates labor, increases the force of contractions, and helps contract the uterine muscle after delivery of the placenta."[102]

Blue cohosh seeds are large and borne atop the foliage; a thin layer of blue flesh covers them, hence the name blue cohosh. The blue draws the eye and lends the plant an attractiveness in fruit that it does not necessarily gain from its inconspicuous flowers. There is an oft-repeated claim that these seeds can be used to make a coffee-like beverage. I have not tried to do so, and as most plants bear well under a dozen seeds, blue cohosh coffee will probably remain a curiosity at best.

Ceanothus americanus • New Jersey Tea

Tea and fire.

New Jersey tea is a multi-stemmed shrub that rarely exceeds three feet or so in height. Its showy white-blooming panicles are ornamental and are attractive to a wide variety of small pollinators.

It is found growing in glades and barrens, on bluffs, as well as on sunny forest edges, usually in somewhat rich soils. It is likely that its growth and reproduction are significantly benefitted by wildfire (and intentional burning).

A passage about western species of *Ceanothus* (the center of diversity for the genus is in California) in the textbook *Forest Ecosystems* is suggestive of the penchant of the genus to rise phoenix-like from the seedbank after fire: "[C]lear-cutting and broadcast-burning old-growth Douglas Fir forests, some up to eight hundred years old, often leads to a thick cover of nitrogen-fixing shrubs in the genus *Ceanothus*. Seeds of these species have persisted in the soil since the previous early succession, hundreds of years ago, and are triggered to germinate by the heat of the broadcast burn."[103]

I often wonder whether the many species of now-rare or locally extirpated plants (of many different genera) might someday spontaneously return to our much-diminished forests here in my home state of New Jersey. Whether there are seeds just below the duff, waiting to be stirred into growth by some (magic?) trigger. In the case of many of our forest herbs, the answer may be no. Species like trilliums, trout lily, and Solomon's seal do not appear to have long-lived seed.[104] In the case of *Ceanothus*, however, the species might lie in wait for a significant disturbance event for close to a millenium!

The data on *Ceanothus americanus* seems to suggest a similar lifeway to that of its western congeners. One study found it to be about twice as frequent on burned vs. unburned oak woodlands in central New York. Likewise, it is described as a "conspicuous dominant" among prairie grasses where fires are frequent.[105]

New Jersey tea (*Ceanothus americanus*).

New Jersey tea foliage produces a delicious tea that is reminiscent of true tea but without the caffeine. In my experience, a fair amount of foliage is needed to prepare a strong brew. Kelly Kindscher suggests that the leaves are best "dried in the shade after being harvested from plants in full bloom in late spring."[106] It was used as a substitute for black tea during the Revolutionary War. New Jersey tea is ripe for consideration as a locally-produced specialty tea.

Herbalists often refer to New Jersey tea as red root and use it as a lymphatic tonic, to increase vascular flow, reverse spleen enlargement, and drain cysts, water-logged lymph nodes,[107] and swollen glands, including swollen prostate.[108] Herbalist Stephen Harrod Buhner writes, "Red root is an important herb in many disease conditions in that it helps facilitate clearing of dead cellular tissue from the lymph system. When the immune system is responding to acute conditions or the onset of disease, as white blood cells kill bacterial and viral pathogens they are taken to the lymph system for disposal. If the lymph system clears out dead material rapidly the healing process is enhanced, sometimes dramatically."[109]

Root nodules on *Ceanothus* host *Frankia* bacteria that fix atmospheric nitrogen; it is one of few non-leguminous species in our native flora to do so. A related species, *Ceanothus herbaceus*, enters our region as a rare species, primarily in barrens in New York State. It can be used in similar fashion to *C. americanus*.

Celtis occidentalis • Hackberry
A thin shell of sugar cloaks the fruit of this tough, polymorphic tree.

Hackberry takes two forms, as a modest-sized, somewhat straggling tree, and as an impressive and distinctive canopy species. The latter form is achieved by trees growing in rich deep soils, usually along rivers. Here, they develop a highly patterned bark composed of warty ridges, the appearance of which is much more haphazard in young and small individuals. Much smaller, craggier trees are found in open uplands, especially on ridges and bluffs. There is some serious taxonomic confusion about the identity of these versus the possible species dwarf hackberry, *Celtis tenuifolia.*

Frequently overlooked as an edible, hackberry bears a fruit with a very sweet and tasty fleshy covering over a large seed. Unfortunately, this sweet edible layer is quite thin. Usage of the pounded fruit, presumably including the seed, is documented among the Dakota, Meskwaki, Omaha, Pawnee, and Kiowa, in porridges, cakes, and for seasoning meat.[110] Forager Sam Thayer discusses eating the fruits whole, cracking the seeds between his teeth, and has high praise for their flavor and nutritional value.[111] When I try I feel like my teeth will crack first. I relegate this food to an occasional sweet treat, but proper processing and further innovation could make it into more of a staple. Thayer covers a variety of preparation methods in his book *Nature's Garden.*

Hackberry is eaten by frugivorous birds and mammals, and is also the larval host plant for the hackberry emperor butterfly. From a restoration perspective, it is a tough, versatile tree that deserves more use. It is "tolerant of salt, drought, alkaline soil, concrete debris ... and moderately tolerant of flooding or saturated soil for up to 25% of the growing season," according to botanist Margaret Gargiullo.[112]

Cercis canadensis • Redbud

Pea family tree.

Arrive in the Appalachians in mid-spring and you'll be treated to roadsides lined with redbud, flowering in deep pink in tandem with the white blooms of dogwood, its close associate. Redbud is a small tree that thrives along well-drained edges and in the understory of open woods.

Redbud's pink blooms and young pea-like pods, are consumed raw, cooked, and pickled, and are said to contain high levels of antioxidants. The young leaves are also edible.[113]

Chamaenerion angustifolium = *Epilobium angustifolium* • Fireweed

Wildflower turns any landscape into a National Geographic photo, and is edible too.

This stunning wildflower once ranged across the state I live in, from the Pinelands in the south to the mountains in the north. It is now essentially gone, despite its simple requirement for sunny areas with sandy or humus-rich soil. What happened here?

Head up to Maine and you'll see this plant as a roadside weed. Look for any picture of Alaska with wildflowers and you might be looking at acres-wide drifts of it.

Easier to find northward, fireweed may have diminished in the southern part of its range due to the lack of fire in the landscape, the intensity of deer-browse levels, competition from invasive species, or some other mysterious cocktail of unfortunate influences. It is not alone in its quiet ebbing away, but is joined by other formerly common species ranging from wild lupine to Indian paintbrush to fringed gentian — all spectacular wildflowers that have simply retreated from view in much of our region.

If you're looking to reintroduce fireweed in an area where it is no longer common, try to obtain seeds or plants from as local a source as possible. Like many native species, fireweed exhibits significant variability and different ploidy levels (the number of sets of chromosomes) across its range. Fireweed from Alaska introduced into New Jersey might do fine, but it won't honor the millenia of divergence and hard evolutionary work done by eastern ecotypes.

Fireweed is used as a tea, and its tender shoots, peeled, are used as a cooked or raw vegetable.

Chelone glabra • Turtlehead

Late season bloomer and balm.

Turtlehead is a late-blooming wildflower with a nearly closed corolla that is adapted for bumblebee pollination. It can be found in both sunny and heavily shaded wet habitats, often on the margins of forested streams.

Herbalists sometimes refer to it as "balmony," and herbalist David Hoffman describes it as a "tonic on the whole digestive and absorptive system," stimulating the secretion of digestive enzymes, and used for gallstones, constipation, jaundice, colic, gall bladder inflammation, and liver problems.[114] David Winston corroborates many of these uses, and adds that "it is especially useful for people with poor fat metabolism, usually accompanied by gas, nausea, belching and a chronically sluggish bowel."[115]

Turtlehead contains iridoid glycosides which deter insect feeding on the plant tissues. These chemicals are found in a variety of medicinal plants, including *Harpagophytum procumbens* (Devil's Claw), where these bitter constituents are considered an anti-inflammatory.[116]

Chenopodium berlandieri and spp. • Goosefoot
Amaranthus retroflexus • Redroot Amaranth

Native annuals.

Before farms and gardens and tillers and bulldozers and empty lots and cracks in sidewalks, what was the indigenous niche of weeds? Where do we find the combination of bare soils, abundant sunshine and other resources that powers the rapid growth and quick, effusive seed production of these transient species? Certainly flooded and scoured riverbanks support diverse weed communities, and necessary conditions can also be found in the excavated soils around groundhog and badger burrows, in the wallows of bison, even on the bare soils of anthills. Just as flooded soils support weeds, so too do burned soils and glacially scoured soils and primary successions after volcanic eruptions, tornados, and earthquakes. In some ways you can think of these species as being synergists with natural disasters and the clawings of wild beasts. In this way, the transition of weeds to being the camp followers, and bulldozer and tiller followers, seems pretty, well, natural.

Many domesticated plants across the world are derived from weeds — logical considering their tolerance for disturbed conditions, their rapid growth, and their positive response to high fertility conditions. Some weeds may have proven their companionability in village refuse heaps, gradually transitioning to the type of deliberately selected and sowed cultigens we associate with domestication.

One of the early eastern North American domesticates was a *Chenopodium* (*C. berlandieri*), part of a crop kit with sumpweed/marshelder (*Iva annua*), little barley (*Hordeum pusillum*), erect knotweed (*Polygonum erectum*), and maygrass (*Phalaris caroliniana*) that was developed in the East up to 4,000 years ago and is known only from the archaeological record.[117] These primarily floodplain species can still be found as wild types, but not in their domesticated forms.

I'm familiar with the goosefoots through consuming lamb's quarters (*C. album*), a common weed that comes up in our vegetable garden every year (and often is the best harvest we get from the garden). *Chenopodium berlandieri* is reputedly very similar. I've found other species in the wild, including *Chenopodium simplex*, a rare species in my state that I've never tasted. Based on my experience with lamb's quarters, Chenopods make a delicious cooked green, hearty and rich and akin to spinach. The seeds can also be ground and dried and used as a flour or meal. The familiar grain quinoa is a domesticated *Chenopodium* from South America. In Mexico, *C. berlandieri* is harvested twice: early for greens, later in the season for seeds, which can be ground into a nutritious flour, significantly higher in protein and fat than corn. The greens are high in calcium and vitamins A and C.[118]

M. Kat Anderson writes that Indigenous peoples in what is now California would sow the seeds of a local *Chenopodium* species after deliberately burning the landscape. This would stabilize soils and provide food while other desirable species responded to the burn by resprouting.

As we create disturbances in our restoration practice, perhaps we should consider *Chenopodiums* and other edible weeds to be our allies in the transitional landscape and sow them as cover crops as the California Indigenous have done.

Amaranths have a similar history as both weeds and prized domesticates. In North America, they are mainly regarded as weeds, though in Central America and other areas of the tropics amaranth species are still cultivated. One species of amaranth, *A. palmeri,* has even developed glyphosate herbicide resistance and is bedeviling conventional farming approaches.

Like the *Chenopodiums,* amaranths have delicious leaves used as cooked greens, and are a pseudocereal (an edible grain that is not a grass). Amaranth grains are notably high in minerals and vitamins, and can be consumed as a meal or popped like miniature popcorn.

Most amaranths in our flora are adventive from farther south and west, or from Eurasia. Redroot amaranth (*Amaranthus retroflexus*) is considered to be a native species in the Northeast.

Claytonia virginica • Spring Beauty

Fairy spuds.

This is a small but cheerful wildflower, harbinger of the end of winter and the renewal of the growing season. Spring beauty has white blooms striped with pink pollinator landing strips, borne atop grass-like foliage. Its flowers open when temperatures are just over 50°F, supplying pollen and nectar to small native bees and flies that become active at those temperatures in early spring.[119] It can form extensive swards in moist woodlands and open fields, spreading by a proliferation of small underground tubers. These are edible and somewhat resemble potatoes in flavor. The foliage and flowers are also edible.[120] Though their size prevents them from being a staple food, this species is still an excellent early season pollinator support, a beautiful herb ... and maybe a seasonal specialty food or attractive garnish.

Spring beauty is one of the few native wildflowers that is highly successful in lawn/turf settings, along with common blue violet.

Collinsonia canadensis • Stoneroot

Unusual mint with strong medicine.

Stoneroot has one of my favorite wild fragrances — something like a mix of citrus and burnt cannabis. If anyone distils it for a men's perfume, count me in. It is a mint, but unlike many in the family, its aromatics are concentrated in the flowers, and the foliage lacks fragrance. It also differs from many other mints in its shade tolerance — stoneroot is a bona fide forest denizen, though it appreciates edges and openings and can grow quite large with moisture and sun. Its flowers are arranged in an open panicle, yellow, fragrant, and two-lipped in the fashion of many mints.

The root is notoriously hard and dense, hence the common name. It is also known as horsebalm and richweed, the latter likely an allusion to the rich soils in which it is found, together with species like black cohosh and purple node Joe Pye weed. As with those plants, it is part of the mesic forest's medicine chest. It is used by herbalists for congestion and constriction in conditions like laryngitis and bronchitis, and also as an astringent for varicosities and hemorrhoids. Its roots contain rosmarinic acid, an antioxidant sometimes used as a preservative.[121] Traditional Cherokee uses include for swollen breasts and sore throat, according to herbalist David Winston.[122]

Comptonia peregrina • Sweetfern

Fern-like aromatic shrub for infusions.

This is a low shrub — usually no more than waist high — with unusual pinnate foliage resembling that found on some ferns. The only member of its genus, sweetfern has a unique appearance as well as a sweet, terpene-laden fragrance. It is brewed for tea, used topically for poison ivy and other skin rashes, and as an antimicrobial.[123] It bears small quantities of edible nutlets, and its immature male catkins are edible as well.

The species can be planted to stabilize steep dry banks such as road cuts. It thrives in low-nutrient soils, handles drought and salt, and is a nitrogen fixer, one of few non-leguminous nitrogen fixing species, along with its relative bayberry, and New Jersey tea.

Stoneroot (*Collinsonia canadensis*).

Coptis trifolia • Goldthread

Tiny creeping herb is a source of powerful alkaloids.

Goldthread is a diminutive evergreen herb often found creeping among mosses in cold swamps. It has distinctive yellow, threadlike rhizomes that contain the alkaloids berberine and coptisine. Alkaloids in goldthread (as well as *Coptis chinensis*, a Chinese relative of goldthread with a similar chemical profile) have demonstrated anticancer effects via apoptosis induction, cell-cycle arrest, and anti-inflammatory activity.

Indigenous peoples use goldthread rhizome for the treatment of infections, poor digestion, stomach cramps, stomach worms, and mouth sores.[124] These uses are in keeping with uses for berberine known from goldenseal (*Hydrastis canadensis*, see below), particularly its digestive use as a bitter to stimulate digestion, and its anti-biotic properties. I would be remiss in not mentioning that the widespread invasive shrub Japanese barberry also contains berberine in its golden yellow, bitter roots, and can be used for some of the same pur-poses as goldthread and goldenseal, which are vulnerable to overharvesting. However, not all herbal uses are reducible to chemi-cal constituents, and goldthread has a character of its own.

Cornus amomum = Swida amoma • Silky Dogwood

Red shoots easy roots.

This is a rapidly growing, thicket-forming shrub in moist to wet soil with ample sun-shine. Its reddish stems are striated with gray, and possess a lanky charm.

While not edible, this species possesses several functional characteristics. It can be established easily from stem cuttings, which are most easily stuck in soil towards the end of the dormant season. It can serve as rapid cover or for bank stabilization. Also,

its fruits are drupes that are high in lipids and help birds prepare for fall migration. Silky dogwood also has a history of usage by Indigenous peoples for both crafts (such as basket weaving featuring its reddish stems) as well as an element of smoking mix-tures admixed with tobacco and bearberry (*Arctostaphylos uva-ursi,* see above).

Red osier dogwood (*Swida sericea = Cornus sericea*) has similar uses and habitat preferences. In my area it is less common than *S. amoma*, having a preference for calcareous wetlands. Its stems are more consistently red than silky dogwood.

Corylus americana • American Hazelnut
Corylus cornuta • Beaked Hazelnut

Native haze 'nuff said.

American hazelnut is a versatile shrub, found in our area on moist forest edges, but also recorded as a native component of midwestern oak savanna and open wood-land communities. It is a multi-stemmed shrub that grows in girth by sending up new stems from a central bole as it grows, often becoming as wide as it is tall. New growth and foliage are covered in a low bristle of glandular hairs. This character helps distinguish American hazelnut from its close relative, beaked hazelnut, which has twigs and leaf stems that are infre-quently bristly.

American hazelnut bears dangling clusters of two to five nuts, each nut enveloped in two downy bracts. Beaked hazelnut, by contrast, encases its hazelnuts in a long involucre that terminates in a narrow tube — something like a beak on a badly drawn cartoon duck.

Why the sudden uptick in botanical lingo here? Because the beak of beaked hazelnut is covered in bristly hairs that shed if handled with bare hands, and can lodge in your fingers like splinters of fiberglass. It's a good idea to be able to differentiate the two species, as American hazelnut can be handled much more readily. Incidentally, this characteristic is shared with its distant cousin, hop hornbeam, whose bracts I found painful to handle when I've collected the fruits for seed.

In my area, beaked hazelnut overlaps in range with American hazelnut. Generally the former is a species of more northerly or high-elevation habitats, radiating south along the Appalachian chain, whereas American hazelnut is well-represented in the Piedmont as well as the tallgrass prairie region.

In addition to ease of harvest, American hazelnut generally has larger nuts. These fruits can still be small, even compared to the store-bought filberts that originate from a Eurasian species, *Corylus avellana.*

American hazelnut (*Corylus americana*).

Among the advantages of our native species is that it is resistant to the Eastern filbert blight, which afflicts the European species, and it is host for at least 20 native lepidopteran species. American hazelnut is also an early spring bloomer and a very showy species in fall foliage.

Most importantly, the fruits of our native hazelnuts are delicious raw, in baked goods, or processed for flours and oil. Indigenous peoples are reported to have preferred the nuts in their "milk" stage, when they are softer and sweeter than mature or dry nuts.[125] Hazelnuts bear abundantly and provide a significant source of quality proteins and carbohydrates.

At least two genetically distinct individuals should be used in plantings to ensure fruiting.[126]

Crataegus spp. • Hawthorns

Heart protector.

The Celts held an erotic spring festival known as Beltane, full of fire and ecstatic dancing and lovers in the forest at night. The blooming of hawthorn marked the beginning of the celebrations.[127] Hawthorn can betoken renewal and healing. It is also a protector, a thicket shrub armed with long, sharp thorns. It is targeted by those who wield saws, axes, and plows, but is deemed a defender of the small and wild things, such as the songbird who shelters from a predator in its twisted, thorny boughs, or the fairy folk of European

tradition, of whom herbalist Sean Donahue writes:

> The old stories tell us that when the world became overrun by men who wielded swords and cut the earth with iron ploughs, the people who taught our ancestors to sing the songs that called the fruit trees to blossom disappeared beneath the hills. At the gateways to their world they planted Hawthorns to tangle and repel the brutish and unwary, but nourish the hearts of those who grieved for lost worlds.

Hawthorn is used by herbalists to support the heart, both in an emotional and physical sense. Hyperin, a form of quercetin found in hawthorns, is a cardioprotective and vascular antioxidant, and anti-inflammatory agent.[128] Hawthorn is a protector of the heart, strengthening and toning blood vessels and the heart muscle itself, and is also used for anxiety, depression, and grief. Herbal preparations feature the flowers, foliage, branch tips, and fruits.[129] Cherokee use the bark tea to "give good circulation."[130]

The division of the genus *Crataegus* into species is fraught for taxonomists and a subject of general confusion for everyone else, due to high levels of variability within described species. Hawthorn identification relies on characteristics from flowers

Hawthorn species (*Crataegus* sp.).

or fruit and sometimes the corroborating evidence of both may be of help. Molecular evidence suggests that *Crataegus* is closely related to *Amelanchier*, another genus of edible woody plants in the Rosaceae that can be hard to cut into species.[131]

The species that I most often encounter include *Crataegus pruinosa, Crataegus crus-galli,* and *C. punctata. Crataegus macrosperma* is also described as common in my region. Some species have more tasty fruit than others. Twice I've found individuals with soft, sweet, uniformly red fruit that is juicy and delicious. These may have been *Crataegus succulenta*, an uncommon species that ranges throughout the Northeast. I collected seed in hopes of having a specimen to identify as well as stock

for propagation, but none germinated. This is typical of *Crataegus*, which are extremely difficult to germinate from seed, due to an impermeable seed coat. Master nurseryman William Cullina suggests grinding the seeds down with "a few minutes of sandpapering" before sowing.[132]

One-flowered hawthorn (*Crataegus uniflora*) is a distinctively short (less than three feet tall) hawthorn species found along the coastal plain and in the lower Piedmont,[133] primarily from New Jersey southward. It has a rounded, egg-shaped leaf unlike many of the hawthorn species in our region, which have sharp lobes. Like other hawthorns, it has thorns and small, orange-red fruit somewhat reminiscent of tiny apples.

Cunila origanoides • Dittany

Diminutive Mediterranean-style mint.

This small, almost shrubby herb is the clos-est species in our flora to a Mediterranean mint like oregano, thyme, or marjoram. It contains essential oils such as thymol, methyl-carvacrol, limonene, and pinene, which are also featured in varying degrees in

those species. Its flavor profile is similar to the native wild bergamot, albeit less spicy.

In late summer, dittany blooms with sprays of purple blooms that support bumblebee populations in the woodland environments where dittany thrives.

The species ranges primarily in the mid-Atlantic, barely entering New York State at its northern extreme. It is primarily found in open woods and glades, and usu-ally in rich soils. During botanical survey work, I often consider it an indicator for unusual plants and interesting geology.

Dioscorea villosa • Wild Yam Root

Heart-leaved vine, medicinal root.

Wild yam is an herbaceous vine growing from a hard underground rhizome. The young leaves are heart-shaped and finely wrought with arcuate (bow-like) veins. These remind me of some Art Deco finery, maybe a soap dish. They twine and sprawl in the forest understory and along edges. Because they are deciduous they don't cause undue distress to the shrubs they climb over.

This plant is favored in Southern folk medicine. Herbalist Phyllis Light recom-mends its use as a liver cleanser for its ability to move bile and reduce "fatty liver." [134] It also was used in the chemical syn-thesis of birth control. However, despite perceptions to the contrary, "the steroidal saponins in the Yams do not act as 'hor-mone precursors' in the body." [135] Rather, it

Wild yam root (*Dioscorea villosa*).

is antispasmodic and anti-inflammatory.[136] Its steroidal saponins also show promising anticancer and antifungal activity, and are hepato- and cardioprotective.[137] According to David Winston, traditional Cherokee uses include for heart, intestinal, and menstrual pain.[138]

Diospyros virginiana • American Persimmon

Fruit of the gods. If you harvest it at the right time.

Persimmons ripen late in the season. Often I am collecting them off of boughs that have already shed their leaves. The black-barked persimmon trees, naked of leaves but laden with orange fruits, fit in well with the general Halloween aesthetic.

And it's no use looking for persimmons too early. Many who have tasted an under-ripe persimmon resist ever tasting one again. Their astringency goes beyond the nuances of flavor, poor or otherwise: astringency binds and precipitates salivary proteins making the mouth feel dry and puckered.

A fully ripe persimmon, soft-fleshed and deep-hued, is another matter and worth the wait. It's a fitting closer for the wild fruit season that begins with wild strawberries, courses through *Rubus* (blackberries and raspberries), *Prunus* (cherries and plums) and various Ericaceae (blueberries and huckleberries), goes tropical for a moment with pawpaws, and draws down with *Viburnums* like blackhaw and nannyberry.

Honey-sweet, fragrant as an apricot, smooth like jam; no wonder persimmon's Latin genus name, *Diospyros*, means "fruit of the gods."

One of the first native plants I tried to propagate was American persimmon. The prior fall I had found a few straggly trees at the edge of the wet meadow near our cottage. I was drawn to the dark, alligator-skin-like bark and then astonished to find small, blotchy orange fruit hanging from jagged twigs. Persimmons! I have no recollection if I knew that native persimmons existed when I first found those trees. I am still impressed when I find a plentiful stand of these fruit bearers and am reminded how delicious they are.

Those first trees I met were in the Piedmont, over diabase rock, in characteristic edge-of-the-forest habitat. Persimmons are well-represented in a variety of habitats and soils, though always cherishing light and edges. They become rarer north of the glacial moraine and in higher elevation mountain habitats. Their greatest prevalence in the Northeast and mid-Atlantic is along the coastal plain, on sandy ecotones. They are also strongly associated with riverine habitats.

Though usually a small tree growing in colonial stands something like sassafras, some individuals can grow to impressive size. For example, the South Carolina state record measured at 132 feet tall in 1995.[139] Meanwhile, in the Outer Banks and Virginia

American persimmon (*Diospyros virginiana*) looking good but not yet ripe.

That's more like it!

Beach area, I've collected delicious fruits from trees that were shorter than I am.

The American persimmon fruit bears some explaining. Typically soft and sticky when ripe, this is not a fruit that ships well or is easy to present for sale. Often it reaches full ripeness in October and beyond, making it a valuable extender for the native fruit season. I've sometimes picked ripe persimmons off of snowy ground, their skin turning blackish purple and the insides a perfect orange hue and honey-sweet flavor.

While we usually consume our persimmons fresh, Cherokee women and other Indigenous gatherers collected and dried persimmons, sometimes kneading them into cakes and pemmican.[140] Others have used them to prepare anything from ice cream to beer. When processing them in bulk it is sometimes a challenge dealing with the remnant astringency that even ripe persimmons can have. Wild persimmons can be very variable in this regard, and a number of cultivars exist that are selected for lesser astringency or firmer flesh.

Persimmons are frequently mentioned in early explorers' accounts of Indigenous husbandry of native fruit trees. For example, Pedro Mártir de Anglería's account of the Carolina or Georgia coast Indigenous peoples in the early 1500s states: "They cultivate orchards, and they have many types of fragrant plants and they like to cultivate gardens; in the orchards they cultivate trees: in particular one that is called *corito*, that produces a delicious fruit." According to historian William Doolittle, one scholar suggests that *corito* was probably persimmon; another fruit described was probably a native plum.[141]

Persimmon is a very flexible species in terms of siting. It grows in a wide variety of soils and is somewhat tolerant of flooding and compaction.[142] Persimmons that are cut or damaged often respond by sprouting prolifically along their root systems, and persimmons can be aggressively colonial in certain circumstances. They are not shade tolerant or competitive with other trees.

Some have conjectured that persimmon's wide range in the wild may be attributable to deliberate selection and dispersal by humans. It is certainly well-attested in the archaeological record: "Historically, [persimmon] was a common component of the Native American diet throughout the southeastern Unites States ... Archaeobotanical data from three Late Archaic period (300–800 BC) sites in the Lower Mississippi Valley found persimmon to be the most ubiquitous fruit crop used ... Paleobotanical analysis of refuse pits from two Mississippian (900–1700 AD) sites identified *D. virginiana* in 78–80% of pits."[143]

It seems that persimmons have been selected for and managed in a variety of ways, on a spectrum from encouragement and protection to outright orcharding by Indigenous peoples.

In addition to being delicious for us, persimmons are of high wildlife value for mammals and birds. Often in the fall one can find scat of foxes, raccoons, and other mammals with persimmon's flat seeds clearly embedded in it.

Southern herbalists have used persimmon root bark, leaves, and unripe fruit as astringent agents in addressing thrush, diarrhea, sore throat, and ulcers. Appalachian folk herbalist Tommie Bass suggested that "if you have stomach ulcers, why, you can make a tea from the bark of the roots and it will kill the ulcer graveyard dead! It makes a real sweet tasting tea and won't pucker you up ..."[144]

Dirca palustris • **Leatherwood**
Plastic and pliable shrub.

Dirca palustris, known as leatherwood or wiccopy, is a widespread but unevenly distributed shrub species with extraordinarily flexible stems. It looks not unlike spicebush (similar leaves and immature drupes) and occupies similar habitats, but it is extremely bendable, like a rubber version of a shrub.

It is this pliability that allows its traditional use in basketry, and as cordage in a variety of applications, even in the manufacture of bowstrings.

Leatherwood "is a deciduous understory shrub of patchy distribution throughout eastern North America ... In its broad range, *D. palustris* generally occurs infrequently but can be locally abundant ... It is found in rich, moist woods, on stream banks and forested bluffs, and often favors north- or east- facing slopes."[145] It is sometimes considered to be a species of old growth (or primary) forests. Slow growth rates have been reported. One author reports that stems two inches in diameter may be more than a century old.[146]

Leatherwood is considered to be highly deer resistant. Limitations to its distribution may be due to the destruction of old growth forest habitat, dispersal limitations, its reputedly slow growth, or a combination of these and unknown factors.

Equisetum arvense • Field Horsetail

Ancient lineage persists as silica-rich scouring rush.

Horsetails are an oddity, an exception, but they were not always so. Three hundred and fifty million years ago, this group of spore-bearing plants towered over the Earth in giant forests. Their high stature may have been enabled by the high silica content in their shoots, a feature still found in the genus *Equisetum*. While relatively tall horsetails are still found in Central and South America (reaching upwards of 20 feet tall on occasion), our local field horsetail is a more modest but still stout member of the genus.

Despite a reduced stature, field horsetail retains a notable ability to spread by its deep horizontal rhizomes. It can be very effective at soil stabilization in moist conditions. *Equisetum* species are mineral-rich plants, bio-accumulators of certain elements in the soil. They have been used for remediation of toxic metals in the soil, which are pulled into plant tissues; field horsetail has even been used as an indicator species for gold mining.[147]

Their high mineral content, and specifically their high silica content, has led to some interesting traditional uses. Ethnobotanist Melvin Gilmore, discussing Indigenous uses, reports: "There is a considerable deposit of silica in the tissues of these plants, and the stems are corrugated. Because of these properties, they are used for finely abrasive material for polishing objects, just as emery paper is used by White people."[148] Some foragers also report the use of young field horsetail shoots as a wild edible. I do not have experience with this usage and would advise careful identification, attention to any toxicity in the soils where it is collected for food, and research into any adverse reactions to eating the species in quantity.

Eupatorium perfoliatum • Boneset

Stem perfoliates leaf on this strong medicine plant.

Boneset is an herb with a strong historical pedigree and increasing scientific evidence in favor of its potency as an antiviral.

Traditionally used for fevers, typhoid pneumonia, malaria, and other serious conditions, recent research points to its use against influenza by protecting cells from infection by inhibiting viral attachment to host cells.[149] Herbalists take alternating chills and fever as an indication for utilizing boneset. Stephen Harrod Buhner, in his book *Herbal Antivirals*, says boneset is used to treat "influenza, dengue fever, malaria, all viral infections with intermittent fever (hot, then cold, then hot, then...)."[150]

A 2018 paper points to a potential hazard of boneset — the presence of pyrrolizidine alkaloids[151] which could be carcinogenic or otherwise toxic. These alkaloids were found in both water decoctions and alcohol extractions of boneset. Pyrrolizidine alkaloids (PAs) are produced by a variety of plant families and serve to defend against herbivory. About 50% of PAs are considered to be hepatotoxic in humans. PAs are found in a number of important medicinal plants, most notably comfrey, causing much controversy among herbalists.

Herbalist Matthew Alfs suggests that "boneset ... is perfectly safe to use, as long as the proper dosage and cautionary guidelines are observed." In his book *Edible and Medicinal Plants of the Midwest*, he lists numerous uses of boneset, including compelling evidence of its utility as an immune system stimulant, an antibacterial and anti-inflammatory agent, an appetitie stimulant (like many bitter herbs), and as possessing cytotoxicity against cancer cells.[152]

Boneset bears white blooming heads on an upright stem that has an unusual leaf pattern. The opposite leaves are fused and so appear to be perfoliated by the stem. It is an excellent pollinator species attracting a diversity of smaller and larger beetles, bees, flies, wasps, and butterflies. Boneset blooms in late summer.

Eutrochium: The Joe Pye Weeds
Gravel roots.

While other Joe Pye species favor marshy conditions, **purple node Joe Pye** (*Eutrochium purpureum*) is a denizen of mature forests, usually found on rich, stony slopes. It is a tall wildflower with whorled leaves and light pink flowers, a bit smaller and more subdued in bloom than its glorious relative, hollowstem Joe Pye. Herbalists call it gravel root, and this double meaning indicates both its habitat preferences as well as its herbal usage in clearing kidney stones. It is also used in formulas for back pain and other musculoskeletal issues, as well as for "low-grade, putrefactive fever" and Crohn's disease.[153] It is somewhat unclear which species of Joe Pye herbalists are using; uses for several species are lumped in herbal accounts under *Eupatorium purpureum*.[154] A good diagnostic characteristic for this species is the concentrated purple spot at each node (as opposed to purple spots and blotches irregularly patterned across the stem in *E. fistulosum* and *E. maculatum*, see below).

Hollow stem Joe Pye weed (*Eutrochium fistulosum*) is a mammoth of a flower, with colossal flowerheads and a stature of up to 12 feet. It prefers moist to wet soils in sunny conditions, and can grow

Hollow stem Joe Pye weed (*Eutrochium fistulosum*).

successfully among tall shrubs in thickets due to its stature. Here on our farm, we have it growing in among numerous tall edible shrubs for extra diversity and beauty.

Its smaller relative **spotted Joe Pye weed** (*E. maculatum*) also favors moist, sunny sites. Spotted Joe Pye is somewhat shorter in stature (usually topping off at four to six feet tall), and has a more flat-topped inflorescence that is darker pink in hue. Unlike hollowstem Joe Pye, which has glaucous stems (covered in a white powdery bloom, like blueberries), the stems of spotted Joe Pye are covered in a thin coat of fine hairs. Both species can feature purple spotting on the stems.

As with boneset, above, there is active research into the pyrrolizidine alkaloid content of *Eutrochium* species. Potential users of the herb should read up on the latest research and consult with herbal professionals before deciding on appropriate uses and doses.

Fagus grandifolia • American Beech
An elder tree of the lowland forest.

Beech is a dominant canopy tree of mature bottomland and mesic forests. Older trees have a high lofted canopy and smooth, pale grey bark. Beech bears delicious nuts, but often the nuts that have fallen to the ground and not been consumed by rodents are not viable, and lack nutmeat. I've never managed to gather beechnuts in quantity, but open-grown trees with low-spreading branches would be the best candidates for a nut harvest.

Unfortunately, beech is being destroyed through part of its range by a blight, beech bark disease (*Nectria* spp.), a complex of fungal pathogens, and the introduced beech scale (*Cryptococcus fagisuga*).[155] A small percentage of wild beech trees are thought to be resistant to this disease complex. Any plantings should probably be done with plant material derived from resistant individuals.

Like other masting trees such as hickories, oaks, and American chestnut, beech is a critical food source for wildlife populations.

Various Indigenous cultures use beech leaves as a dermatological aid and create an oil from nutmeats.[156] Beechnut oil is popular in Europe (derived from *Fagus sylvatica*). The nuts are deliciously sweet; at one time, they were commonly available in grocery stores in New England.[157]

In northern hardwood forests, beech is an important component, codominant with sugar maple. Both are shade tolerant, later successional species, often recruiting in the understory below other, pioneering species such as oaks, ash, red maple, and birch. Beech leaf litter is low in nutrients and is slow to decay.[158] In combination with the dense shade cast by beech trees (especially younger ones with branches still low on the trunk), this can cause the ground layer under beech trees to be sparse.

American beech (*Fagus grandifolia*).

Fragaria virginiana • Wild Strawberry

Heartberry.

Picture this. I'm singing the praises of one of our tastiest wild fruit plants to a potential nursery customer, both for flavor and as an amazing groundcover. "Oh, I already have those in my lawn," comes the oft-repeated reply. But no, you don't! You're harboring an imposter.

Because there's *wild* strawberry, and there's *Indian* strawberry. The latter is also known as mock strawberry or false strawberry.

Take the golf-ball-sized strawberries at the supermarket, distill down all the flavor into something that size of your pinky nail, make it ten times better, and you get a *wild* strawberry.

Wild strawberry (*Fragaria virginiana*).

Take a wild strawberry, screw around with the leaves, make the flower yellow instead of white, and make the fruit a barely edible nugget of insipid flesh, and you have *Indian* strawberry.

Got it?

Wild strawberry is a low, delectable creeper, a native species deeply held in the hearts of those who know it. It grows in sunny spots in dry fields, woodland edges, and trailsides. It is not uncommon but its extent may be diminished by the lack of fire in the landscape. Early colonist Roger Williams (quoted by Kelly Kindscher) gives a charming account of wild strawberry dating to 1643 that mentions Indigenous cultivation of the species: "The berry is the wonder of all the fruits growing naturally in these parts. It is of itself excellent, so that one of the chiefest doctors in England was wont to say, that God could have made, but God never did make, a better berry. In some parts where the Indians have planted, I have many times seen as many as would fill a good ship, within few miles compass." [159]

Wild strawberry figures prominently in Cherokee mythology, in the archetypal relationship saga "The Origin of Strawberries" transcribed by James Mooney from Cherokee sources, and found other places as well.

Indian strawberry (*Duchesnea indica*), on the other hand, actually is from India. It occupies ruderal (weedy, disturbed) places in the landscape — lawns, gardens, flood-plains. More or less the same kind of places you'd find the well-known creeping weed gill-over-the-ground. To repeat, it tastes like nothing good and is barely edible.

Some people have tried the latter, the imposter strawberry, and decided that wild strawberries are tasteless. Go check some dry meadows for wild strawberry this year. The species fruits in middle June in northern New Jersey where I live.

In addition to being delicious, wild strawberry is an ideal permaculture groundcover. It seeks this way and that beneath established perennials, suppressing annual weeds while playing nicely with the shoot vegetables, herbs, trees, and shrubs you've planted in the same area.

The Lenni Lenape call this plant heartberry. Once we've got the taxonomy straight between wild strawberry and its imposter, maybe we can work on changing its designation back to this beautiful and fitting praise name.

Gaultheria hispidula •
Creeping Wintergreen

It has moxie.

This is a small, creeping woody subshrub native to evergreen-dominated northern swamps. Its fruits and foliage have a mild wintergreen flavor. It is used for fresh eating, preserves, and tea. It is sometimes know as "moxie plum." A wintergreen flavor such as that found in this species was an ingredient for the soft drink known as "Moxie," which gave us the colloquial phrase that someone "has moxie" — lots of energy or courage.[160] It is the host plant for the bog fritillary butterfly.

Gaultheria procumbens •
Wintergreen

Winter cheer and a cool flavor.

With dark evergreen leaves and bright red fruit, this diminutive groundcover adds cheer and color to the winter woods, where it is found among other heaths and deciduous oaks as well as fellow evergreens like pines and mountain laurel. Its fruits (just a tad sweet) are imbued with wintergreen flavor, the key ingredient in birch beer and many contemporary root beers. The leaves are somewhat less sweet but offer a cool wintergreen chew with some tannic overtones — perfect for tea as well. Despite its short stature (usually below six inches tall), this is a technically a shrub, spreading to form colonies with miniature woody stolons.

The plant's wintergreen flavor is due to the phytochemical methyl salicylate, from which a closely related chemical was derived to produce aspirin. Wintergreen is mildly pain-relieving (analgesic) and anti-inflammatory.[161]

Like many heaths, wintergreen requires acidic, sandy soils to thrive. Those looking for wintergreen flavor in more rich, mesic habitats should consider black birch (*Betula lenta*), above.

Gaylussacia baccata •
Black Huckleberry
Gaylussacia frondosa •
Blue Huckleberry

Not a blueberry! Maybe even better.

How is it that blueberries are so widely known and appreciated and huckleberries so obscure? They occupy similar habitat, look similar, and have similar fruit, so perhaps it is a case of one being subsumed into the identity of the other. In the wild it is certainly possible to be snacking on lowbush blueberry and move right on to huckleberry without noticing it.

I'd like to make a case for recognizing the huckleberries. Black huckleberry has dark, glossy black fruit (lacking the powdery bloom that makes many blueberries look blue), grey-black stems, and a somewhat greater stature than the lowbush blueberries. I find huckleberries to be even tastier than blueberries, with a bit more sweetness relative to acidity, and a pleasant seedy crunch due to its slightly larger seeds. In addition to fabulous fruit, huckleberry shrubs have vivid red flowers, and stunning scarlet fall color, especially dramatic when seen in the large colonies they are wont to grow in.

As with some of the other heaths, black huckleberry is deer resistant.

Blue huckleberry, also called dangleberry, is a spreading, medium-height shrub with an affinity for sandy, moist soils. Its fruits are blue and somewhat larger than black huckleberry. The taste is similar to but not always as sweet as black huckleberry in my experience. Botanists Richard Porcher and Douglas Rayner are fans, stating, "Huckleberry makes one of the most luscious of desserts, being juicy with a rich, spicy, sweet flavor." [162]

Black huckleberry (*Gaylussacia baccata*).

Geranium maculatum • Wild Geranium

Early season pollinator support and astringent herb.

Wild geranium is a beautiful woodland wildflower, thriving on edges and in canopy gaps. Blooming right after the height of spring ephemerals, it signals the transition to larger blooms and welcomes the first butterflies. It supports a number of different types of native bee including *Andrena distans*, which is a specialist on wild geranium.[163] Its roots are strongly astringent. As such, it is used to tone mucous membranes and address conditions such as diarrhea and other "excessive discharges."[164] Appalachian folk herbalist Tommie Bass used wild geranium roots for internal bleeding, hemorrhoids, and bleeding ulcers. According to his apprentice Darryl Patton, "[wild geranium] is ... very useful for treating intestinal bacteria and protozoa. It pulls the moisture from the invader, which then dehydrates, collapses, and dies."[165]

Geum rivale • Chocolate Root

Herb makes chocolate-like beverage.

Chocolate root? This low herb of sunny seepages, bogs, and fens is seldom remarked on, but it has a root that makes a delicious, rich beverage somewhat analogous to chocolate. I won't exaggerate the resemblance, because chocolate is magical stuff (remember carob? blechh), but this native herb stands on its own as far as its distinctive, rich flavor is concerned. The only issue I've encountered is that the roots are fairly small so it requires a fair amount of plant material to make more than a cup or two of hot chocolate root.

The flowers of this species stand out due their deep red-purple calyx, and the plumose radiating red styles persisting on the seeds make for an unusual and attractive show.

Algonquins use the plant for "the spitting of blood." Other Indigenous uses include for diarrhea, fever, dysentery, coughs, and colds.[166]

Geum rivale is used in European folk medicine as an astringent and antiseptic. Recent research has confirmed the presence of significant quantities of antioxidant polyphenols.[167]

Gleditsia triacanthos • Honey Locust

Sweet pods, thorny trunk.

This leguminous tree produces seed pods filled with a sweet pulp surrounding the seeds. The pulp can contain 20–35% sugar.[168] The pulp is best used when the pods are young, and edibility for humans may vary. It is being used as a perennial feed source for hogs and other animals in integrated livestock farms.

While its current range is in the central United States, this species may have been dispersed by now-extinct megafauna and has not been able to radiate into its former (pre-Ice Age) range, which was wider. However, humans have stepped in (as with other possible megafauna-dispersed fruits like pawpaw, persimmon, and osage orange) and have spread honey locust to a wider range than considered to be its "natural" range. Some evidence points towards dispersal and cultivation of this tree by Cherokee in the past.[169] Honey locust is also fairly common in contemporary landscaping, where thornless varieties are valued as street trees because they leave little leaf litter. The thorns on wild-type individuals are quite impressive, branched and very sharp.

Native American medicinal uses of the plant are widespread. According to herbalist James Duke, "Honey locust pods are a folk remedy for dyspepsia and measles among the Cherokee. The bark tea is used for whooping cough. Delaware Indians used the bark for blood disorders and coughs, the Fox for colds, fevers, measles, and smallpox."[170]

Hamamelis virginiana • Witch Hazel
Medicine for assholes.

Witch hazel is a tall shrub with a high canopy, each stem seeking sun almost heedless of the others, so that sometimes it sprawls and stems and branches intertwine in pursuit of a small light gap. It is most noted for its flowers, which though spidery and wan, bloom in late October into November when little else blooms.

Witch hazel had some folkloric use in water-witching (dowsing), something I did as a kid on rambles with my dad but am unsure I could pull off now. Mostly, it is noted as an astringent, especially for topical use, and it is one of relatively few herbal

Honey locust (*Gleditsia triacanthos*).

medicines you could find in fluorescent-lit drugstores even in the dark ages before echinacea and so on.

We clip twigs into small finger-length portions, soak them in a mason jar of vodka for a month, and use that for weepy rashes, like poison ivy. It is particularly, amazingly good for fixing a torn-up itchy butthole after a few days of diarrhea, and hemorrhoids in general.

The homemade tincture, black and aromatic, beats the pants off of the nasty but weak rubbing alcohol stuff at drugstores, and ought to be an item of cottage industry for every wildcrafter with a woodlot.

Hedeoma pulegioides •
American Pennyroyal

Tiny mint with powerful volatile oils.

This is a small, inconspicuous annual plant, more noticeable by fragrance than appearance. I often find this little herb while hiking by brushing against or stepping on it, which releases a piercing, cold, minty aroma. Its primary use is as an absolutely delicious, refreshingly cooling tea. It can be used as an ingredient in natural mosquito repellents. Like its relative, European pennyroyal, this plant has been used as an abortifacient[171] in high concentration. As a precaution, pennyroyal tea should not be drunk by pregnant women who wish to remain that way.

Pennyroyal's small seeds recruit readily in bare soils that have received some

disturbance, and on uplands scoured free of leaf litter by winds. It is primarily found in sunny, dry areas such as canopy gaps, trail edges, and glades. American pennyroyal is pollinated by small bees including Halictid bees, little carpenter bees, and the dagger bee, *Calliopsis andreniformis.*[172]

Helianthus annuus •
Annual Sunflower

Sunflower seeds.

There are about 70 species of *Helianthus*, all native to the Americas (only three of which are native to South America). Of this quintessentially American genus, two have been widely utilized as food plants: Jerusalem artichoke (see below), and this, the annual sunflower.

Sunflower is well-known for its familiar oversized flower heads and abundant black seeds. It is an Indigenous domesticate which may date back 3,000 years — seeds found in Salts Cave in Kentucky are larger than those of the modern wild type of the species, suggesting deliberate breeding and cultivation.

Sunflower seeds are used by Indigenous peoples across North America in a variety of ways, including as an oil, or ground to a meal with which to make breads and cakes.[173] Lewis and Clark's journal from 1805 records that the Indigenous people of the Missouri "parch the seed and then pound them between two smooth stones untill they reduce it to a fine meal. [T]o this they sometimes mearly add a portion of water and drink it in that state, or add a sufficient quantity of marrow grease to reduce it to the consistency of common dough and eate it in that manner. [T]he last composition I think much best and have eat it in that state heartily and think it a pallateable dish." [174]

Though sunflower domestication was known across the continent, the wild-type sunflower (or various feral admixes) is also common, and persists as an annual, "weedy" species in disturbed sunny areas. These too were used for food by many Indigenous peoples. Early ethnobotanist Melvin Gilmore reports that "Arikara people told me that though the seeds of the wild are inferior in size they are superior in flavor." [175]

Native to the East or not? It is believed that wild type *Helianthus annuus* may have originated on the Colorado Plateau. It is considered native only as far east as the Mississippi. However ... sunflower figures into the Iroquois creation myth. The wild species is well represented in every state in the East. How you treat this in terms of native species restoration is up to you. Some may choose to utilize traditional cultigens, others the wild seed, others may eschew it as a waif in our region. Those who subscribe to the latter view may have to examine the possibility that other native species were also dispersed by Indigenous people and decide whether human agency can be

considered "natural" in their philosophical perspective and whether that is important. It is not a simple question. If Indigenous peoples' agency is "natural," this quickly leads to difficult questions about labeling contemporary imported plants (including invasive species) as non-native or exotic, because, after all, aren't humans animals and our movement of plant species "natural"?

Does thinking this way undercut the usefulness of the concept of "native" or "natural"? In my mind it doesn't, but rather points to a deeper reality, which is that all such overarching categories bleed at the edges, and that there is no such thing as a universal truth. Instead philosophies and tactics are situational, relative, and strategic — everything is context dependent. Just like in ecology, where cardinal flower is strange without hummingbird, but together they are perfect. The utility of a label such as "invasive" occurs within a context of diversity collapse, habitat destruction, deer overabundance, and the simultaneous invasion of hundreds of other species. This context makes distinctions such as "exotic," "non-native," or "invasive" critical in understanding the global degradation of ecosystems.

Those in favor of utilizing sunflower can consider some of its functional traits. Sunflower (at least the domesticated form) is a heavy feeder that can deplete soils of excess fertility if harvested and removed from the site. This might help with reducing excess nutrient availability in post-agricultural sites that would otherwise be heavily prone to weeds. Sunflower is also widely used in bioremediation, and has the ability to uptake the heavy metals nickel, copper, arsenic, lead, and cadmium into its tissues.[176, 177] One lab study documents its potential in phytoremediation of the antibiotics tetracycline and oxytetracycline.[178]

Agricultural literature details potential allelopathic effects of sunflower on certain crop species, particularly if its residues are incorporated into soils.[179, 180] This could be a benefit or a drawback in ecological restoration practice depending on whether weeds or desirable species are suppressed. One dissertation from 1968 on oldfield succession in Oklahoma and Kansas shows sunflower inhibiting germination and/or seedling growth of a variety of weedy field-colonizing species, including horseweed, black-eyed Susan, crabgrass, amaranth, and Japanese brome.[181] A more recent study used sunflower leaf aqueous extract as a natural herbicide and found reduction of *Rumex dentatus* and *Chenopodium album*.[182]

Helianthus tuberosus • Jerusalem Artichoke

Beautiful but aggressive spreader with edible tubers that nourish our digestive biome.

This is a large, colonial sunflower species that grows in rich soils. It is a species with a history of deliberate cultivation by Indigenous peoples and may still be found at the margins of crop fields that were once Indigenous fields. Its presumed native range centers in the Ohio and Mississippi River valleys, so populations in the Northeast very likely originated from introduction by Indigenous peoples.

Unlike the familiar annual sunflower (*Helianthus annuus*), this is a perennial species. It bears smaller but very prolific flowers. Its seeds are small for food use, but it produces edible tubers — enlarged sections of root that are rich in the carbohydrate inulin. This dietary fiber feeds the beneficial bacteria in our stomach biota and confers "benefits upon host well-being and health"[183] — the host being us and the benefit sometimes profound. However, take it slow at first — a side effect of feeding so many stomach organisms on inulin, an indigestible starch, is often profuse gas. Longer cooking times and other processing methods can mitigate this. Because the inulin is ultimately healthful for our system, the best approach may simply be to start with smaller amounts and work up to larger portions over time. Note that the inulin in Jerusalem artichoke tubers is naturally converted to fructose in the spring as the plant prepares to fuel new growth.[184]

This species exists in both wild and domesticated forms, with the latter having larger (albeit less flavorful) tubers. The domesticated variety can sometimes be more productive than potatoes. There is some contention as to whether Indigenous peoples domesticated this species; it is clear, however, that it was dispersed outside its original range by people, and thus encouraged by a variety of peoples in eastern North America.[185]

Jerusalem artichoke is a strongly spreading and persistent species. Tubers can radiate several feet away from the original plant in a season or two, spawning large colonies. Goldfinches and other seed-eating birds relish the nutritious seeds, close kin to the annual sunflower used for birdseed.

Hierochloe odorata • Sweetgrass
A sweet reminder of reciprocity.

This modestly statured aromatic grass has extensive uses among Indigenous peoples as a ceremonial incense or smudge. It is used in basketry by many Indigenous peoples including those within the Iroquois Confederacy of Nations. I have never seen this species in the wild, partially because its habitat in my area largely consists of salt and brackish marshes, areas where I haven't ever done fieldwork and am only an occasional visitor. More importantly, it is also declining rangewide because it thrives best when harvested according to traditional methods utilized by Indigenous peoples. This reciprocal relationship between a plant and people is woven through the book *Braiding Sweetgrass* by Robin Wall Kimmerer. I told a bit of the story back in Chapter 6. Rather than paraphrasing further, I'll encourage you to read her book which is without a doubt one of the best books ever written about plants.

Hydrastis canadensis • Goldenseal
Treasure of the eastern forest.

Goldenseal is a low herb with palmately lobed leaves vaguely reminiscent of maple or fig leaves. Spreading by rhizomes, it resembles mayapple in its colonial habit.

Its namesake yellow rhizomes are imbued with the alkaloid berberine, a pleasantly powerful bitter that has demonstrated antibacterial, antifungal, anti-inflammatory, anticarcinogenic, antiamoebic, and antimalarial activity.[186] Uses by herbalists are myriad. Darryl Patton summarizes its folk uses as an antibiotic, blood purifier, and "strong liver medicine" and as a remedy for styes, ulcers, skin disease, and inflamed tissues.[187]

I've used goldenseal as a mucous membrane tonic during sinus infections, as an antiseptic for wounds, and as a digestive bitter; like other bitters it stimulates the production of digestive enzymes and aids the digestion of fatty foods.[188]

Goldenseal is the only plant in its genus — a unique herbal treasure of the great eastern woods. Millions of goldenseal rhizomes are harvested from wild populations yearly, threatening its survival in the wild. There is a strong need for organically cultivated sources of the herb. As with other wild medicinals, it seems to best express its phytochemistry when grown in a manner that closely replicates its natural setting. When grown with artificial fertilizer it contains lower levels of medically beneficial alkaloids.

This is logical, since compounds like berberine can be a response to environmental stresses and the creation of these compounds may not be triggered under artificial, pampered growing conditions.

Goldenseal has an affinity for mafic and calcareous geologies, with soils high in magnesium or calcium. Some years ago I found one of the only known populations of goldenseal in New Jersey, and it was on diabase geology near other rarities such as wild comfrey (*Cynoglossum virginianum*) and pennywort (*Obolaria virginica*).

Hydrophyllum virginianum • Virginia Waterleaf

Woodland groundcover with edible leaves and bumblebee flowers.

Virginia waterleaf is a seldom-discussed native plant that makes a terrific groundcover for part shade and shaded areas. It has beautifully patterned leaves which emerge early in spring, attractive flowers in midspring that support native bees, and it forms a dense, low mass of foliage. It is edible as a green both raw and cooked. A recipe from the Menominee uses the foliage "wilted in maple syrup vinegar, simmered, and boiled

Goldenseal (*Hydrastis canadensis*).

in fresh water with pork and fine meal." [189] Young foliage is best, and waterleaf puts on new growth at several times in the season, unlike some native woodlanders whose

Virginia waterleaf (*Hydrophyllum virginianum*).

growth is predetermined the prior season as dormant buds form.

Waterleaf tends to grow near streams and rivers, or in deep soils, sharing rich habitats with wood nettle, wild ginger, spikenard, linden, sugar maple, and the like.

Hypericum punctatum • Spotted St. John's Wort

Native substitute for the European species?

This modest wildflower has the open yellow flowers typical of St. John's worts, but they are small, less than half an inch across. It grows in moist and mesic sunny conditions, persisting as a perennial where it is does not face competition from taller species.

It is a potent source of hypericin and other constituents that have caused its European cousin, St. John's wort (*H. perfoliatum*) to be acclaimed by herbalists and widely utilized in alternative medicine. Herbalists praise St. John's wort for its beneficial effects on nerve pain and muscle spasms, whereas alternative medicine has promoted hypericin as an antidepressant.

It is unclear whether the native species can act as a substitute for the European species and mirror all its uses. A recent study found that constituents in the native *Hypericum punctatum* (and related species) inhibited growth of Gram-positive bacteria, which are biofilm-embedded species, including *Staphylococcus epidermidis,* and clinical methicillin-resistant *Staphylococcus aureus* (MRSA).[190] Further exploration of the medicinal uses of this common and versatile herb are warranted.

Ilex vomitoria • Yaupon

Might give you a buzz but won't make you puke.

Yaupon is an extremely widespread shrub native to the coastal plain from Virginia south. It has small, leathery evergreen foliage, bright red fruits, and is readily cultivated.

Several early European accounts described cultivation of yaupon in the 18th century; Adair writes "the Indians transplant, and are extremely fond of it"; Bartram describes it moved out of its coastal plain habitat to the mountains by the Cherokee. He notes a grove tended by Indigenous people, who called it "the beloved tree, and are very careful to keep it pruned and cultivated."[191] Archaeological evidence in the form of dippers or cups suggest that yaupon was utilized as early as 1000 B.C.E., and cultivated beyond its native range, as "an important element in cultures of the Eastern Woodlands."[192]

Yaupon makes a tasty, caffeinated infusion. It is the only caffeine-bearing species in our flora. I like to prepare the leaves by dry roasting them for about ten minutes sometime prior to brewing them. Fans of yerba mate, a South American beverage brewed from an *Ilex* species, will recognize a relative in yaupon, though its flavor is different.

Some accounts described the beverage historically brewed from yaupon as "black drink," and claim that it was used in Indigenous rituals to induce vomiting — hence its specific epithet. At least one apocryphal tale to the contrary suggests that colonial-era tea merchants paid off some botanists to give it a distasteful reputation and nomenclature because its domestic popularity threatened imported tea sales. I can't weigh in on either side, but can attest that normal doses brewed a tasteful amount are not in the least bit emetic. Those searching for a native caffeinated alternative to coffee or tea, with much lighter environmental impact, need look no further.

Juglans nigra • Black Walnut

Nuts and syrup from a fast-growing tree renowned for beautiful wood.

Black walnut is a tree of paradoxes. A fast grower, it also produces hard, dense wood of incredible beauty. A known allelopath, it produces chemicals that discourage the growth of other plant species around it. However, it also leafs out late, drops leaves early, and casts fairly thin shade, making it an ideal companion for spring- and fall-blooming plants.

Black walnut is a canopy-height floodplain species that was frequently planted on old farm fields (as an investment, in the old days), and naturalizes in disturbed areas. Its nuts are distributed by squirrels and others, and are also considered a delicacy by some people, especially for the flavor they impart to baked goods. The fruits are high in unsaturated fats, including omega-3 fatty acids, and the liver-protective antioxidant glutathione.[193] A highly regarded Cherokee dish mixed corn grits with black walnut or hickory nutmeats.[194]

At least as good as the nuts (much better in my opinion) is the smoky dark syrup that can be derived (in the same manner as maple syrup) from the sap of black walnut.

Black walnut is used by herbalists (especially in the South) for ringworm, fungal infections, internal parasites, canker sores, ulcers, psoriasis, and thrush.[195] Southern Appalachian herbalist Phyllis Light, who grew up in a folk tradition that highly values the plant, recommends black walnut as an antifungal, antiviral, and antibacterial agent, to expel parasites, and to support liver function including some aspects of detoxification.[196]

Black walnut is a versatile, disturbance-tolerant tree that often recruits in waste areas and neglected edges. It can handle soil compaction, flooding, salt, and drought to a moderate degree.[197] It is easily propagated. The chemical juglone, contained in leaves, bark, husks, and roots of black walnut, can suppress the growth of some other plant species (including younger black walnuts). From a restoration perspective, this can offer an opportunity to create heterogeneity in the landscape, as certain species may be selected for under black walnut.

Kin to black walnut is the **butternut** (*Juglans cinerea*), which wins high praise for the flavor of its nuts. It is a fast-growing, shade-intolerant species. Unfortunately, butternut is succumbing to butternut canker, a fungal disease, across most of its range. Many butternuts scattered on old farmsteads are hybrids with a related Japanese species. Black walnut may be threatened as well, by the emerging pathogen thousand cankers disease, currently found primarily in the West but with at least one eastern occurrence.

Juniperus communis • Common Juniper

Prickly oldfield shrub with spicy cones.

Common juniper is a shrub found in clearings and old fields; its sharp needles are protective against grazing livestock and small stands of it can pepper old pastures. Its fruit-like cones are eaten by rodents and songbirds, and also provide a botanical used to flavor gin, as well as a spice for various savory dishes. Note that various

Common juniper (*Juniperus communis*).

sources caution against its use,[198] or prolonged use, citing toxicity issues.[199]

The species is widely used in Indigenous traditions as a cold remedy, cough medicine, pulmonary aid, pain reliever, and ceremonial medicine.[200]

Laportea canadensis • Wood Nettle
Urtica gracilis • American Stinging Nettle

Richly nutritive edible greens growing in lush but stinging colonies.

Wood nettle is a tall, colonial herb of rich, deep soils, such as those found along rivers and streams, and in rich mountain coves. It resembles its relative, stinging nettle, with its stinging hairs. However, the native wood nettle has larger leaves, which furthermore are arranged alternately (as opposed to opposite) along the stem.

Wood nettle complements stinging nettle well, in that it arises later in the growing season (emerging after most spring frosts) and offers tender shoots and foliage after stinging nettle has become coarse. Also, its larger leaves are more tender and more savory as well. I assume that its nutritional content is on a par with stinging nettle (which is well-known for its nutritional value and high protein level), but lack data to substantiate this. Presuming this to be the case, wood nettle can also be used to create herbal tea for people and compost tea for plants, with a similar effect to stinging nettles.

Nettles broadly speaking are also fiber plants. Of all the indigenous nettles used for fiber, including **smooth nettle** (*Boehemeria cylindrica*), a wetland species, and **stinging nettle** (*Urtica gracilis*, see below), a riparian species closely resembling the European stinging nettle, wood nettle was the most popular for use by Indigenous peoples. Archaeologists have discovered wood nettle fabrics produced by Indignenous cultures in Kentucky, Tennessee, and Ohio, and more contemporary use is known among the Iroquois, Ojibwe, Sauk, and Fox peoples.[201] Indigenous uses for wood

Wood nettle (*Laportea canadensis*).

nettle also include the use of smashed roots for various psychological and physical ailments, and a decoction of the plant taken for fever.[202]

Wood nettle can be established along partly shaded edges, especially given the presence of some soil moisture, or naturalized along waterways and swales. It is a capable spreader in moist shaded habitats and creates dense patches.

American stinging nettle (*Urtica gracilis*) is so similar to European stinging nettle that it was long considered a subspecies (*Urtica dioica* ssp. *gracilis*). These two species are used interchangeably for food, fiber, and medicinal uses. Of particular relevance for the native species are long-standing Indigenous uses as a counter-irritant for muscle and joint pain, as an agent to relieve dermatological itches, rashes, and sores, and to facilitate childbirth.[203] For identification purposes, the native species is differentiated from the European by being mostly monoecious, by having shorter leaf teeth, and importantly, by having fewer stinging hairs on its stems and on its leaves. I have identified this species in the field occasionally but don't have a sense for the percentage of wild stinging nettles in our region that are native or not. Alan Weakley's *Flora of the Southeastern United States* shows both species as common in the mid-Atlantic and declining southwards. Its habitat includes bottomland forests over rich geology.[204]

Larix laricina • **Tamarack**

Bog medicine.

Just as seeing a sycamore at a distance is a great indicator for a hidden river or stream, seeing a tamarack always thrills me as it suggests the presence of a bog, with all of its miniature plants, rare species, and mats of multihued sphagnum mosses.

Tamarack is a deciduous conifer that grows as far north as any tree species can survive. It is not adapted to shade, but rather thrives in open, moist to wet sites; it is frequently an indicator for northern bogs though is not entirely confined to them.

Like many conifers, tamarack's young leaves provide a pleasant nibble; they can also be brewed into a mild tea rich in vitamin C. Tamarack's inner bark contains an impressive pharmacopeia. Among these are arabinogalactans, which, according to botanist Arthur Haines, are immune system modulators, prebiotics, preventative in liver and colon cancer, and antidiabetic.[205]

Lepidium virginicum • **Virginia Pepperweed**

Spicy little weed.

This species is a weedy little herb that thrives in disturbed sandy areas. It is used for its mildly spiced foliage, and for its spicy seedpods, both of which have a mustardy pepperiness to them. It's a good green to toss in a salad or use as a spice when some heat is called for.

Pepperweed can be added to restoration seed mixes for initial soil stabilization as it is a rapidly germinating annual (or biennial).

Lindera benzoin • **Spicebush**

Aromatic shrub of moist soils.

Spicebush is a colonial shrub that bears small lemon-yellow flowers in spring. Female plants produce deep red glistening drupes in the fall that are an important food for migrating thrushes. The entire plant is suffused with a complex of aromatic constituents that can be described as sweet and slightly citrus-like.

Spicebush is a forest species adapted to shady conditions and modest nutrient availability, producing long-lived, high-quality root tissue as compared to many weedy or invasive woody plant species.[206] Spicebush responds well to enhanced light availability and can become dominant following logging and other disturbances that remove canopy trees.

The crushed dried fruit of spicebush can be used as a culinary spice, and its twigs can be used for herbal teas and steams. Because we harvest spicebush seeds in order to grow the seedlings in our nursery, we have a surfeit of the fleshy fruit coating. We have dried and powdered this and made an excellent spicebush-flavored ice cream, which is a bit like vanilla and cinnamon but is unique.

Species in the genus *Lindera* are used across continents for medicinal purposes.[207]

Spicebush is underutilized at present but has a history of use by Indigenous peoples across its range. Traditional uses include use as a steam for pain relief and to induce sweating, and use for menstrual pains and delayed onset of menstruation.[208] Herbalist David Winston reports use of the herb for "colds, flus, coughs, nausea, indigestion, croup, flatulence, and amenorrhea." [209]

Spicebush is a host plant for spicebush swallowtail and eastern tiger swallowtail butterfly larvae.

Liquidambar styraciflua • Sweetgum
Aromatic amber resin from star-leaved tree.

Sweetgum is a canopy tree that thrives in sandy, moist to wet soils. It has palmate leaves that are almost star-shaped, and spiny, spherical "gumball" seed capsules that are familiar to many because the species is often planted as a street tree.

Sweetgum exudes an aromatic resin when cut. The resin was called *xochiocotzoquahuitl* by the Aztecs, for whom it was a trade item and an ingredient in resin incense. The resin is the namesake for the tree's common name as well as its Latin designation — *Liquidambar* (amber liquid) *styraciflua* (aromatic resin flow).

Sweetgum ranges from New York State as far south as Nicaragua, and it has been used traditionally by cultures throughout its range. Sweetgum's resin has been used in smoking mixtures, as a chewing gum, and in soaps and cosmetics. Dr. James

Duke's Ethnobotanical Database lists over 70 traditional uses ranging from headache and toothache remedy, to herpes and halitosis remedy, to wound-healer. Contemporary North American herbal use is scant, suggesting this as a species ripe for rediscovery. Tommie Bass recommended the use of the bark as a "wonderful tonic" and used a syrup of the resin for coughs and colds. His student Darryl Patton recalls Bass's use of it for the flu, and comments that sweetgum contains shikimic acid, the active constituent in the prescription drug Tamiflu.[210]

Sweetgum's range is primarily on the coastal plain. It is tolerant of saturated soil for up to 75% of the year, and is moderately salt tolerant, but does not grow well when shaded.[211] While it is often found in the canopy of alluvial forests and swamps, it can also readily colonize fields and logged areas. Sweetgum bark is relished by beavers, and small birds such as finches harvest seeds from its spiny seed capsules.[212]

Liriodendron tulipifera • Tuliptree
Canopy or bust.

This incredibly exotic tree can be so commonplace that I sometimes forget to notice how remarkable it is. If it were a rarity we'd all moon over its enormous, magnoliid blooms licked with orange, yellow, and green flames, its highly unusual cat-faced leaves (with apex indented rather than pointed out as with other leaves), and its incredible size. Tuliptree grows straight-trunked towards the sun, is highly apically dominant, and can attain heights of nearly 200 feet tall as an old growth tree.

Tuliptree's easily peeled, almost leathery bark makes excellent bark baskets. Medicinally, it is an astringent, bitter tonic, used by the Cherokee "for periodic fevers, diarrhea, pinworms, as a digestive aid and for rheumatic pain. The decoction is used as a bath for fractures, sprains, hemorrhoids and is applied to snakebites received in dreams," according to herbalist David Winston. Tommie Bass recommended a hot tea decocted from the roots for arthritis, rheumatism, and as a diuretic.[213]

Tuliptree was one of the native trees recommended as an emergency antibiotic and antimalarial during the Civil War, when the Confederacy faced shortages of *Cinchona* bark for malaria treatment. Botanist Francis Porcher compiled a book in 1863 that included Indigenous alternatives for antiseptics and antimalarials, useful in the treatment of soldiers. Among those selected were tuliptree, white oak, and devil's walking stick (*Aralia spinosa*), which were analyzed in a recent lab-based study. The study verified the bioactivity of these recommended species, and found that tuliptree extracts demonstrated activity in inhibiting bacterial growth, quorum sensing, and biofilm production. However, tuliptree branch bark extracts also demonstrated significant mammalian cytotoxicity,

suggesting careful attention to parts extracted and dosage for any herbal usage.[214]

Known as *tsiyu* among the Cherokee, tuliptree is used for fevers and diarrhea, for fractures and sprains,[215] as a digestive aid, and for rheumatism, according to David Winston. The part utilized is the smooth young bark.

Tuliptree flowers harbor a secret treat. The tepals of the flower harbor a divine, honey-like nectar deeply infused with floral

Tuliptree (*Liriodendron tulipifera*).

tones. Sample it by finding a newly opened flower and licking the sweet nectar from the green outer parts. This nectar is also attractive to hummingbirds.

Tuliptree produces numerous wind-borne seeds that blanket adjacent areas after fall and winter winds. The shadecloth on our greenhouses is often peppered with its winged samaras in November. It colonizes forest gaps after logging, burns, and windthrows, and tuliptree seedlings grow rapidly in partial to full sun. Sometimes it is possible to detect selective logging performed decades in the past due to groves of uniformly aged tuliptrees found within otherwise mixed forests. Despite being a "pioneer" tree (some might prefer to call it a healer of gaps), tuliptree can persist as a venerable member of old growth forests. In older age, its thick, corky bark confers some fire resistance, and its high stature keeps it from being shaded out.

Lycopus uniflorus •
Northern Bugleweed

Wetland tuber.

Northern bugleweed is an easily overlooked herb of wetland edges and shores, with small axillary flowers and opposite foliage. The tuberous roots of northern bugleweed species were used steamed or baked as a staple and as a dessert by the Sylix (Okanagan) and Niaka'pamux (Thompson) Indigenous peoples.[216] I've never seen these bugleweeds in the large quantities

necessary for subsistence. If growing them for food, one would need to innovate a way to non-lethally harvest the plant, perhaps by replanting part of the harvested root, to make the initial effort of planting worthwhile from a food utility perspective.

Forager Sam Thayer describes bugleweed tubers as "mild and slightly sweet ... crispy when raw but tender and somewhat starchy when cooked." Though this book is not a foraging guide, I will pass along his caution to carefully discriminate between the roots of bugleweeds and water hemlock, which can be found in the same habitat and has similar looking roots.[217]

Appalachian herbalist Tommie Bass employed two related species medicinally, *L. virginicus* and *L. americanus*. He utilized these species for coughs, bronchitis, and in his nerve tonic formulas.[218]

Maianthemum racemosum •
False Solomon's Seal
Maianthemum stellatum •
Starry Solomon's Plume

Shoots and fruits.

These are two upright to slightly arching herbs with starry plumes at the end of their stems. False Solomon's seal (also known as Solomon's plume) inhabits mesic and mesic-upland forests, and starry Solomon's plume is found in sandy coastal forests and in sandy floodplains. Both herbs have edible shoots in spring before the leaves completely enlarge and unfurl. The fruits, borne in terminal clusters, are also considered edible.[219]

Matteuccia struthiopteris •
Ostrich Fern

Giant prehistoric fern with gourmet fiddleheads.

Ostrich fern is a tall, colonial fern that grows in thick swards, often across large areas. Each ramet (individual in a clone) consists of a cluster of elegant, plume-shaped fronds arising from a thick rhizome. In the spring, each frond unfurls from the rhizome base where it lay deeply curled through the winter.

These unfurling fronds are known as fiddleheads, and those borne by ostrich fern make a delicious, almost meaty, shoot vegetable. This species is relatively easy to establish in moist shade, where it is likely to establish a nearly monospecific sward over time. Ostrich fern fiddleheads are a well-known seasonal vegetable in the Northeast, and locally grown fiddleheads could be a good seasonal cash crop for market farmers. Peak harvest time as well as habitat coincides with wild leeks, offering the possibility of a seasonal package of wild foods with excellent flavor and high market appeal.

A recent four-year study observed that crowns from which 100% of the fiddleheads had been removed "suffered significant decline in the number of fiddleheads produced over the four subsequent years" and suggested that 50% or fewer fiddleheads be harvested from a single plant in a single year.[220] I'd suggest this as a baseline for management of a planted stand. When foraging a wild population, I take only a single fiddlehead from each plant.

Although ferns are not much used now in the herbal pharmacopeia, Woodland Cree used a decoction of the leaf stalk for back pain and to speed expulsion of the afterbirth.[221]

Ostrich fern is found primarily from Pennsylvania and New Jersey northward, becoming most common in northern New England into Canada. It is uncommon or rare in several mid-Atlantic states.

Medeola virginiana •
Indian Cucumber Root

Underground cucumber.

I had always been curious to taste the intriguingly named Indian cucumber root but never found it in sufficiently abundant and healthy colonies to forage it until one day when were hiking in the Appalachian Mountains in North Carolina. We were trudging uphill on one of those hikes where you keep thinking "we must have made it to the top" and then a whole vast stretch of further uphill hiking is revealed. The kid was getting cranky, we were all hot and thirsty, and then ... the trail on either side opened into swards of Indian cucumber root, hundreds in all. We picked up digging sticks and in a minute had dug three white tubers. I still remember how our eyes lit up on sampling these moist, crunchy, sweet rhizomes, very like their namesake except even more perfectly crisp and flavorful.

I had been reticent to try this forest herb because harvest is generally lethal for the plant; one is eating the rhizome from which the individual arises. If searching

for a sustainable harvest option, one *could* harvest a portion of the "cucumber" and replant a bit of rhizome.

Micranthes pensylvanica = Saxifraga pensylvanica • Swamp Saxifrage
Swamp vegetable.

Swamp saxifrage is an herb of wet woods, stream corridors, and swamps, with broad, tongue-like basal leaves and an upright stem bearing tight, starry flowering clusters at the top. The young foliage is edible. The plants are frequently browsed by deer (especially the flowering tops), and I have only tried the species once because of the scarcity of flourishing local populations.

Monarda didyma • Bee Balm
Scarlet beauty with aromatic flowers and foliage, attracts hummingbirds.

A showy mint with truly beautiful scarlet flowers, bee balm's preferred habitat is sunny openings along streams. Its flowers and foliage alike are fragrant with volatile oils. Bee balm's flowers are a choice edible, combining thyme, rose, and citrus-like aromatics. Its red tubular floral corollas attract hummingbirds as well as large butterflies such as swallowtails. Bee balm's leaves makes a very good herbal tea as well as a spice that can be used like oregano or the aforementioned thyme.

Bee balm is used for colds, headache and stomach discomfort.[222] We use bee balm as a relaxing tea to soothe stomach

Bee balm (*Monarda didyma*).

upset due to anxiety and stress. Chemical analysis shows bee balm to have a high percentage of thymol (over 50%) as well as other volatile oils, and lab tests show activity against pathogenic fungi and bacteria (*Escherichia coli, Erwinia amylovora,* and *Candida albicans*), while selectively less inhibitory towards beneficial microorganisms.[223]

Monarda fistulosa • Wild Bergamot
Monarda punctata • Dotted Horsemint

Healers in the meadow.

Wild bergamot is a showy and versatile wildflower in the mint family. Its aromatic principles fall more on the thyme and oregano end of the spectrum rather than the cooling spearmint end. It has a spicy bite, a bit like the heat of a hot mustard (as opposed to chili pepper) but with a lingering tingle.

Herbalist Matthew Wood describes wild bergamot as a hot, pungent nervine and stimulant. Calling the plant by its Indigenous name, "sweet leaf," he describes it as one of the "half dozen ... most important herbs used by the Indians of North America." Wood describes the many uses of the herb in his own practice, from healing burns and septic wounds to calming and toning nerves and releasing passion.[224] At home, we use the herb for burns, tick bites, wounds, colds, and feelings of dizziness.

Wild bergamot makes a fine tea herb, as well as a good oregano substitute for pizza, pasta, and the like. Indigenous peoples used the herb variously as a seasoning for food (especially meat), as a tea, and for cosmetic uses as well, such as a fragrance for the hair. Kelly Kindscher observes that

Wild bergamot (*Monarda fistulosa*).

the young foliage is sweetest for tea, the later leaves make the best oregano substitute, and the mature leaves becoming increasingly spicy.[225] I harvest the young shoots, up to a foot tall or so, sizzle them in a skillet with oil and salt, and find the resulting crispy greens to be particularly delicious.

The flowers of wild bergamot are lavender-colored tubular corollas growing in whorls from a spherical center. If you've never seen a hummingbird moth, find some wild bergamot in bloom and you'll see one of these hovering pollinators, looking like an iridescent flying crayfish, feeding on the nectar. Wild bergamot also attracts numerous butterflies and bumblebees.

Wild bergamot germinates readily and reliably from seed and is a frequent component of meadow and prairie restoration seed mixes. It is often one of the first, and showiest, flowers to bloom on a seeded restoration site. Its habitat preference is for mesic to dry, gravelly openings, meadows, and edges.

Four or five subspecies and varieties of *Monarda fistulosa* are currently recognized, and Indigenous peoples considered the different varieties to be important, with differing uses and medicinal character assigned to each, some being, for example, taller, spicier, ill-smelling, shorter, more oily, or more sweet.[226]

Dotted horsemint (*Monarda punctata*) is a widespread species with a preference for sandy soils in full sun. As such, it is found in meadows and openings throughout the coastal plain (south to Florida), as well as inland into the prairie region. It has purple-spotted creamy flowers subtended by showy, purplish bracts, giving the flower an outsize, exotic look and attracting numerous pollinators.

It has a flavor profile similar to wild bergamot: dry, spicy, and thyme-like. As with the other *Monarda* species, it has numerous medicinal applications based around its volatile oils. My friend Julie Martin of Practical Primitive has some in a jar that's ten years old and is still potent enough to clear your sinuses out with a good whiff.

Morella pensylvanica • Bayberry
Bay by the bay.

This is a tough shrub, tolerant of salt, dry, wet, sand, or clay, and resistant to deer browse. The only thing bayberry doesn't handle well is shade. The sunny dunes of the windswept, salt-sprayed shore are an ideal long-term habitat for it. Bayberry grows in low, rounded mounds, often wider than tall, as it sends out lateral shoots in concentric circles from the original crown, and as sand deposition buries stems.

Brush up against the plant on a hot day, or crush a leaf, and you'll inhale its rich, bay-like aroma, with a brisk overtone of citrus. This is a fragrant spice, usable in stew like the common culinary bay leaf. It makes

Bayberry
(*Morella pensylvanica*).

a very fine tea, with a satisfying tannic bite and a texture like green tea. Bayberry's waxy fruits are also used in the production of an aromatic candle wax.

Bayberry is a nitrogen fixer, associating with *Frankia* bacteria to fix 14 to 28 lbs of nitrogen/acre/year.[227] This is a significant addition in the nitrogen-poor sands of the coastal environment. Bayberry's presence facilitates superior growth from beach-grass, seaside goldenrod, and pitch pine. It has a mild inhibitory affect on little blue-stem and has been documented replacing this native grass in successional fields.[228]

Rutgers University professor Ted Stiles and his colleague Allan Place studied the yellow-rumped warbler, a bird species that can readily be observed feeding on bayberry in late summer into winter. While most animal species find saturated long-chain fatty acids — bayberry wax, for one — to be fairly indigestible, they documented that these warblers readily digest it, with greater than 80% assimilation efficiency.[229] Because this food is available primarily to yellow-rumped warblers, and bayberry fruit tends to persist into the winter, the warbler is able to overwinter well north of other warbler species.

In addition to coastal habitats, bayberry is found inland in a variety of wet and dry landscapes. The related species **wax myrtle** (**Morella cerifera**) has a more southerly distribution and some overlapping uses.

Morus rubra • Red Mulberry
A tree full of berries.

In my area, red mulberry is disappearing fast, possibly as the result of competition from and genetic swamping by the non-native white mulberry. In fact, I am still looking to find a mature wild tree of this species in my state. By all accounts, this is a superior fruit, and forager Sam Thayer describes it as better than the Asiatic species white mulberry, which is a high compliment.

Red mulberry is a versatile forest tree, more shade-tolerant than white mulberry, which grows in weedier circumstances. Red mulberry utilizes its broad leaves as large solar panels to catch and photosynthesize the dappled light in forest understories. Leaves that develop in the sun tend to be more lobed than the broad, simple shade leaves; this lobing helps reduce overheating and transpiration.

Red mulberry bark has a number of uses. Cherokee women wove the bark into floor and wall coverings, and used it for making soft clothing as well.[230] Bark tea was utilized by the Cherokee and Cree, to expel worms, for dysentery, and as a urinary aid.[231]

Several early European accounts that allude to Indigenous peoples' protection, encouragement, or cultivation of native species mention red mulberry. Peter Kalm in the Delaware Valley reported finding red mulberry with sumac, chestnut, and black walnut in "old corn-fields." William Bartram observed red mulberry trees in the "ancient cultivated fields" of central Georgia with persimmon, honey locust, chickasaw plum, black walnut, shagbark hickory, and American beautyberry, and commented that "[t]hough these are natives of the forest, yet they thrive better, and are more faithful, in cultivated plantations." John Smith, in the early 1600s, noted "[b]y the dwellings of the [Indigenous] are some great Mulbery trees" in Virginia, and likewise in New England "many faire high groves of mulberie trees gardens."[232]

Red mulberry tolerates short periods of flooding. It is very high in wildlife value for birds and mammals.

Myrica gale • Sweetgale
A northern bay substitute and aromatic smudge.

Sweetgale is a small, aromatic shrub that grows on shores and in nitrogen-poor bogs, aided by its association with nitrogen-fixing bacteria. It is related to bayberry (*Morella pensylvanica*, see above), and has a sweet, rich aroma. Like bayberry, its foliage can be prepared to make a fine tea or used to flavor stews.

In the first volume of *Ancestral Plants*, botanist Arthur Haines documents medicinal uses of this plant in the treatment of gastrointestinal disorders, the common cold, and as a skin wash for insect bites, poison ivy, and minor wounds. The inner bark and the foliage are utilized depending on application.

Nelumbo lutea • American Lotus

Gigantic aquatic food source with spectacular flowers.

The American lotus features enormous, cupped flowers up to a foot across, the largest of any North American wildflower. It is stunning, unique, and large in every way. Colonies of the plant can cover acres of still riverine habitat, ponds, and lakes. An emergent species, its leaves and blooms emerge from the water while its roots sink deep into the soil.

Like other lotus species, its round leaves are hydrophobic — water will bead up on its surface and run off. It is quite similar to the Asian sacred lotus but has yellow rather than pink flowers.

Pretty much the whole plant is edible, including flowers, shoots, roots, and young seeds. For those familiar with Asian lotus root, this is a similar starchy "vegetable" marked by a beautiful geometric pattern of openings. Forager Green Deane describes the roots as edible raw or cooked, and reminiscent of sweet potato. The leaves can be cooked like spinach, and older leaves used to wrap food. He describes peeled stems as tasting like beets.[233] Porcher and Rayner write, "The ripe seeds [are] parched to loosen the shell, then husked and eaten dry, baked, boiled, or ground to make bread." The seeds are also a valuable wildlife food, especially prized by ducks.[234]

Seeds of American lotus can lay dormant for centuries, awaiting shallow inundated conditions to trigger growth. They might lay in abandoned lakebeds or

American lotus (*Nelumbo lutea*).

forgotten river oxbows awaiting a change in hydrology. Seeds of its close relative, the Asian lotus, have been awoken from dormancy after 1,200 years!

The species is rare in much of the mid-Atlantic, with the core of its range found around the Mississippi River and tributaries. Populations in parts of New England are believed to be introduced and some states treat the species as invasive, for example where it competes with other native riverine species such as wild rice. J. S. Newberry, a sort of proto-ethnobotanist of the 19th century (and the first geologist to explore the Grand Canyon), suggested that several northeastern populations might have been introduced by Indigenous peoples.[235]

Nymphaea odorata • White Pond Lily

Beautiful floating aquatic with edible flower buds.

I was once hired to do a botanical survey on a 37-acre lake in northern New Jersey. The surface of the lake was nearly covered in white water lilies. It was a spectacular sight to behold, thousands of diaphanous white petals geometrically radiating from sun-gold centers, perched on glossy Pacman-shaped leaves, over dark lake waters. Glorious — but hard to paddle through.

This profligate abundance of beauty may also represent an abundance of food. Ojibwe people consume the unopened flower buds.[236] The young leaves are edible

cooked, and some state that the rhizome, or tuberous growths on it, can be rendered into flour or "prepared as a potato substitute."[237]

This species can grow in water at least six feet deep; during the aforementioned lake survey we recorded it in areas with depths of up to nine feet, but it was not necessarily apparent where it was rooted. This species spreads vigorously and provides food for aquatic birds and mammals, habitat for fish and invertebrates, and can cool water temperatures by shading the water surface.[238]

Nyssa sylvatica • Tupelo

Tortoise tree with acidic fruits.

Tupelo, often known as blackgum, "often represents one of the few tree species that predates the original cutting of the forest," according to a monograph by forester Marc D. Abrams. This is due to a historical accident — tupelo is undesirable as a timber tree and was often just left as forests were logged of more valuable species. As a result, individuals of this slow-growing, shade-tolerant species can be found that are centuries old, mixed into stands of much younger trees. The oldest known individual of this species, at 679 years old, qualifies it as one of the longest-living hardwood species in the world, and individuals of "ages of 450 to 550 years are not unusual," according to Abrams.[239]

Tupelo has a wide latitude of habitat tolerances, and is able to grow in almost every habitat type in the East as far north as Maine, and into Canada north of the Great Lakes. It has a great deal of phenotypic plasticity and is often divided into subspecies reflecting regional and habitat variations. It thrives in low-competition environments, and many of the oldest known tupelos are found in swamps. It is also drought tolerant, fire tolerant, and shade tolerant, an unusual trio of attributes for a late successional, long-lived tree species.[240]

This is a distinctive tree, with brilliant red fall color and very horizontal branches. It often grows in small colonies radiating out from a mother tree along its root system. While the bark on young individuals can be somewhat nondescript, over time mature trees develop a wonderfully checkered, snakeskin appearance, with deep furrows describing erratic polygons across their trunks.

For many years, I thought its dark blue fruit was just for birds (it supports over 30 bird species, including the charismatic pileated woodpecker). I knew a close relative in the south, *Nyssa ogeche*, is used for its sour fruit and dubbed "Ogeechee lime." It wasn't until many years into admiring this tree species that a friend suggested I try the fruit. It is pleasingly acidic, puckery like a lemon but equally inviting. Some report its use in preserves. I have not gotten past the trailside nibble phase yet, but could see this fruit being used in place of familiar

Tupelo (*Nyssa sylvatica*) and the author (*Homo sapiens*).

citrus in desserts, beverages, and savory dishes. Like many other largely forgotten wild edibles, it needs a new generation of interpreters and creative cooks to take it under their wings. Note that fruits are borne only on female trees.

In many ways, tupelo reminds me of a tortoise. It is slow growing but extremely long-lived. Its characteristic bark has a tortoise-shell pattern. Given large forest disturbances, it is quickly outpaced by faster-growing species. But a century later, it is still there, plodding along towards the canopy, feeding the birds, and blazing with brief glory every fall. It's a good archetype for a sustainable existence and a symbolic tree that could diversify restored landscapes that are intended to far outlast us.

Oenothera biennis • Evening Primrose

Healthy seeds on a prolific weed.

As with the goosefoots featured above, this is a native weed with edible properties. The primary edible part of evening primrose is the carrot-like taproot. Many report enjoying eating it, but I've always found it gives an odd, scratchy sensation to my throat after ingestion.

The oil of evening primrose seeds contain essential fatty acids including omega-6 fatty acids. Clinical applications include usage in the treatment of dermatitis, mastalgia, diabetic neuropathy, and rheumatoid arthritis.[241]

Evening primrose has large, pretty yellow flowers that are primarily open from dusk to dawn and are pollinated by moths. Butterflies and bees flying early or late in the day or taking advantage of overcast conditions function as backup pollinators.[242]

Evening primrose is a biennial that readily seeds into bare, disturbed soil.

Opuntia humifusa complex • Eastern Prickly Pear

Northeastern cacti.

Most people don't associate the Northeast with cacti, but I've seen this adaptable little cactus everywhere from shale cliffs over the Delaware River to glacially scoured ridgelines to the sandhills of western Wisconsin. Its core habitat is along the shore, where it thrives in the hot sands of coastal back dunes. A versatile species, prickly pear is probably limited in more mesic habitats not by moisture but simply by the fact that a short, sprawling cactus is readily overgrown by other species.

Kept free of competition, this species can form dense beds, which become absolutely glorious when the golden yellow flowers open. Its fruits are edible but small, seedy, and mucilaginous — I've yet to crack the code on really enjoying them. The pads are a good edible, grilled or sauteed and served Mexican style. The pads of *Opuntia macrorhiza* are reportedly boiled to loosen the skin; after its removal the soft insides are fried.[243]

Eastern prickly pear cactus can reproduce by seed — the seedling cacti are tiny, fuzzy, and adorable. They are also slow growing. A faster mode of reproduction (in the wild and for restoration use) is via broken-off pads, which will form roots in contact with soil. After Hurricane Sandy in 2012, prickly pear pads were found washed up along the shore of Rhode Island, a state in which only one population was previously known.[244]

Taxonomy of the *Opuntia humifusa* species complex is … complex. The species *Opuntia humifusa*, *O. cespitosa*, *O. drummondii*, and *O. austrina* are all found along the East Coast according to a recent taxonomic clarification.[245]

Osmorhiza claytonii • Sweet Cicely
Osmorhiza longistylis • Aniseroot
Licorice-like sweetness.

These are two inconspicuous herbs in the carrot family, with ferny compound leaves and small white flowers. They possess a sweet, anise-like flavor, aniseroot in its foliage and root, sweet cicely mainly just in its roots. The flowers and young seeds of each also have this flavor. These parts can be used to flavor a delicious tea, and can be used medicinally for sore throat, coughs, and disturbed digestion. David Winston suggests that medicinal uses (for sweet cicely) overlap to some degree with licorice and astragalus,[246] both powerful herbs with compelling traditional uses. He adds that among the Cherokee, the root is used as a medicine for "increasing strength, weight, and resistance to disease," and that the tea is useful for cold symptoms such as sore throat and dry cough, as well as flu and digestive issues.[247] These overlooked woodlanders are worth further investigation as both a flavoring and a medicine.

Oxalis violacea • Violet Wood Sorrel
A little sourness.

This is a pretty little woodlander, with open, cheery purple blooms over little shamrock leaves. It has a lemony sourness that serves well as a nibble, in a salad, or to add tang to a soup. Its bulbs are also edible.[248] Ecologically, it can hold down a small area of disturbed ground (such as trailsides and borders). It might be used as a secondary species in the restoration of edible mesic forest plants.

Panax quinquefolius • Ginseng
Grandfather of herbs.

If there is a holy trinity of Appalachian medicinal herbs it is ginseng, goldenseal, and black cohosh, each unique to the eastern forest. Of these, ginseng is surely the most legendary, with its humanoid taproot and profound healing abilities.

A long-lived perennial (six decades or more), ginseng was formerly widespread in the rich deciduous forests of the East. However, over a century of intensive harvest, coupled with deer overbrowse in recent decades, has rendered this plant uncommon over much of its former range.

Ginseng (*Panax quinquefolius*) has a long history of unsustainable harvest for commercial trade. In 1860, the state of Minnesota alone exported over 120 tons of dried ginseng roots to China for medicinal use.[249] The first-ever American direct trade with China took place in 1784

American ginseng
(*Panax quinquefolius*).

and featured ginseng prominently among trade items.

Ginseng is a nourishing adaptogen that supports recovery from illness and fatigue, chronic stress, and some auto-immune diseases. Recent studies associate ginseng compounds with anti-aging effects, improved stamina, and anti-inflammatory action after physical stresses.[250]

Cherokee sometimes call ginseng *a'tali-guli,* "the mountain climber."[251] Ginseng relishes the moist soils of sloping, rocky woods with exposures that are shielded from the most intense direct sun.

Passiflora incarnata • Maypops, Purple Passion Flower

Fruit and calming medicine from a vine with spectacular blooms.

Passiflora is a genus with several hundred species in the American tropics. Only maypops and one other species range in the temperate United States. Passion flower has a tropical personality, with a fancily wrought flower and exotic-tasting fruits. It is a vine, rising every spring from dormancy and sending its twining stems across other plants, fences, and any other supports it can find. Passion flower spreads via underground runners, and can be aggressive.

Its native range is from Maryland and Delaware southward. Passion flower tolerates freezing temperatures down to a range of about 0–10°F and thrives in areas with an average frost-free period of 160 to 200 days.

Paleoethnobotanist Kristen Gremillion details an extensive human-plant mutualism between Indigenous peoples and passion flower, stating "it presumably was restricted to relatively open ground beneath canopy gaps (including stream margins, windthrow gaps, and other naturally disturbed patches in the landscape) before disruptive human activity became common within its range." Passion flower is a beneficiary of human disturbance, and remains a common vine of disturbed edges from Virginia southwards, growing on fencerows, along roadsides, and in thickets and vacant lots. It is tolerant of a wide variety of soil types but is not shade tolerant.[252]

Accounts from early 17th and early 18th century Virginia describe maypops growing in Indigenous gardens among the corn, in what could be construed to be a kind of fruit and grain polyculture. According to Gremillion, "[S]ources do indicate that maypops grew abundantly in Indian gardens, that the fruit was consumed, and that the plant was tolerated and possibly encouraged or planted in Historic times."[253]

Maypops have a spectacular flower with showy styles and stamens. Bee pollination is necessary for fruit set.[254] The fruits, also called maypops, are tropical-flavored and sweet with a hearty tang; they are particularly well-regarded as a jelly ingredient. In addition to supporting bees, it is a host plant for several butterfly species.

Passion flower herbage is the source of useful medicine. The foliage and stems of the plant are tinctured to make an herbal remedy of particular effectiveness. I keep *Passiflora incarnata* tincture by my bedside and use a small dose whenever I can't fall asleep due to mild anxiety or being overstimulated. Countless times the last thing I remember before falling asleep is taking my five drops of passion flower tincture. It is a non-narcotic sleep aid and is also used to relieve pains and spasms.[255] Passion flower is also antiviral. Herbalist Phyllis Light recommends its use for herpes and other viral infections.[256]

Phytolacca americana • Pokeweed
Potent weed.

This is a flamboyant, oversized herb, with huge weedy stalks, glossy dark fruit, potent toxins, and a Bollywood color scheme of reds and purples. It is ready to take residence in whatever heap of disturbed earth it can find, whether a tip-up mound in an old growth forest, or a corner of the yard. It germinates from seeds that can remain dormant in the soil for decades.

The young shoots of the plants are a tasty spring edible, and contain calcium, iron, phosphorous, and good quantities of pro-vitamin A and vitamin C.[257] To avoid toxicity, shoots should be harvested while still green (i.e., before turning strongly red). Shoots will generally be six inches tall or shorter during their harvest window.

Pokeweed is antimicrobial and stimulating; both roots and berries are used in herbal practice. The roots are used by some herbalists for respiratory tract infections and other infections. Herbalist David Winston describes Cherokee uses for the plant as including "rheumatism, skin conditions, [and] as a poultice for swollen breasts."[258]

This is a toxic plant and should be eaten and used medicinally with proper precautions. That said, it's interesting how virulent the warnings are about this plant in most contemporary resources, which treat the fruit and other plant parts as toxic. Some Indigenous peoples seem to have been less wary of the plant, and the entry for pokeweed in Moerman's *Native American Food Plants* describes Cherokee uses of the berries: "Crushed berries and sour grapes strained, mixed with sugar and cornmeal, and used as a beverage."

You decide. (Carefully.)

In addition to edible and medicinal uses, the plant yields a gorgeous purple dye from its fruit.

Generally, there is no need to introduce pokeweed to your site, as it gets around quite well just on its own. The fruit is of high value for fall migrant birds and is readily sown about the landscape. Pokeweed germinates readily in response to human disturbances such as burning, clearing, and gardening.

Picea: The Spruces

This bud's for you. And the resin too.

In spring, you'll notice a change in the coloration of the dark, rigid needles of spruce trees. New growing tips emerge, much lighter green and softer than prior years' growth. These pluck off easily and make for an unexpectedly terrific vegetable. They have a bit of citrusy terpene bite, together with an almost meaty richness. They are tasty raw but become superb cooked.

Spruces also offer a versatile product in the form of the resin that exudes from any surface wounds on the tree. This has been used as a salve on wounds and a topical antimicrobial,[259] as a respiratory remedy,[260] as an old time chewing gum, and even an adhesive, especially when admixed with charcoal.

Black spruce (*Picea mariana*) is a relatively small, slow-growing conifer found locally in bogs and cool swamps south to New Jersey and Pennsylvania but becoming most abundant in the challenging conditions of the boreal forest in northern New England. **Red spruce** (*Picea rubens*) is similar, found in cool, moist sites. Both species are shallowly rooted; red spruce is somewhat more shade tolerant. It is also distributed somewhat further south, as it tracks the Appalachians down into North Carolina. **White spruce (*Picea glauca*)** is the most northerly of the group, found primarily in Maine and in the mountains of other New England states, and northward into Canada to the Arctic treeline.

Pinus: **The Pines**

Needles for all your needs.

The pines of our region occupy a breadth of habitats, from sandy coastal wetlands to exposed ridgelines. Generally speaking, they are most competitive in poor soils, where the costs for pines associated with maintaining evergreen foliage are not as significant as the costs for deciduous trees of trying to manufacture new leaves each season in infertile soils or in places with short growing seasons. Many of our pines are also fire adapted, tolerant of or even dependent on fire for regeneration.

While each species has a distinctive ecological niche and personality, the uses of pines are held in common by all the species in our area. Pines offer a wealth of edible and medicinal uses.

The inner bark of pines is edible and nutritive, containing high levels of antioxidants. It increases the potency of vitamin C in the body. The bark can be ground into a flour and admixed with other flours (it has a strong flavor on its own).[261]

The tender shoots of spring are edible, as are young leaves. Both young and older leaves ("needles") can be used to make a flavorful, healthy tea, and I enjoy nibbling on pine needles throughout the year.

Syrup of white pine was an official drug listed in the United States Pharmacopoeia for coughs. Herbalist Darryl Patton suggests making a hot tea of the needles as a gargle for sore throat, or chewing on the pine resin, which makes a sort of gum that used by Indigenous people as well.[262]

Both the pollen cones and the pollen itself make interesting and highly nutritive foods. Young pollen cones are consumed prior to opening, and can be cooked up like little corn on the cob. They are very rich in vitamins (A, B complex, and C) and minerals including calcium, magnesium, phosphorus, iron, manganese, potassium, and sodium.[263]

Pine pollen is increasingly well known for its phytoandrogens, which can boost testosterone levels when taken over a period of weeks. It is also powerfully antioxidant.[264] It can be added to foods or tinctured.

Pine resin (the sticky "sap" one sometimes sees oozing from wounds) is rich in terpenes and can be used as an external antiseptic. Sometimes the inner bark is used as a dressing for wounds and burns. Pine resin, bark, or leaves can be steamed as an inhalant to treat congested sinuses and respiratory complaints associated with colds. Mary Siisip Geniusz notes that the sap is readily harvested from red pines, as attested in its specific epithet *resinosa*.[265]

In addition to edible and medicinal uses, the bark, roots, and resin of pines all find application in the creation of cordage, containers, glue, and other cultural items of Indigenous peoples.

Jack pine (*Pinus banksiana*) is a stunted and hardy occupant of cold, sterile, burnt, windswept, and/or lumbered regions of the North, a pioneer tree once thought to poison the soil around it because it frequented such barren habitats. It is relatively short, relatively short-lived, and tough as nails. It is sometimes considered a nurse tree in the harshest habitats, sheltering species that may succeed it.[266] I've seldom had a chance to meet this tree, but remember its stark beauty on slabs of granite on Great Wass Island in Maine. This is at the southern extreme of its range and is an unusual site in that the Jack pine is fairly long-lived (up to 150 years old) and succeeds itself in multi-age stands.[267] Jack pine is the most fire-adapted of all the boreal conifers. Its reproductive strategies, including serotinous cones (which open from fire), early reproductive maturity, and ability to germinate in mineral soils[268] are all of benefit in areas with regular fire return intervals.

Red pine (*Pinus resinosa*) is a far northern species, at the southern limit of its range in the Acadia region of Maine.

Red pine is a shade-intolerant species that grows quickly in sandy and rocky soils. It is often found in dense stands with foliage only near the top; this limits the accessibility of both needles and young cones, two of the primary edible and medicinal parts for all pine trees.

Pitch pine (*Pinus rigida*) is often a dominant species in pine barrens, though in the absence of fire it can be replaced by oaks and other hardwoods over time. Pitch pine does not compete well on rich sites but finds its niche in landscapes with high fire-return intervals and poor soils. It is also found on ridgelines in the Piedmont and Appalachian Highlands, where excessively drained conditions limit competition. It displays unique fire adaptations including serotinous cones (opening after fire), as well as non-serotinous cones, and Adventitious buds that can resprout after fires. Epicormic sprouting of adventive branches gives this pine a wonderfully shaggy appearance unlike any other. Coupled with its contorted branching, I find this to be among the most interesting-looking species in a genus full of beautiful trees. Pitch pine can be highly variable in form; harsh conditions can modify it into a dwarf tree or even a shrubby mat.[269]

Pitch pine was much used in the historic era for its water-resistant timber, and for its tar and resin. Donald Culross Peattie, that poet laureate of American trees, relates that "Pitch pine knots ... are yet so filled with resin that they resist decay long after the stump has rotted away, and in regions where the tree was abundant, they used to cover the forest floor. Pioneer children were kept at work, stooping and gathering these, day after day. The knots were then tied to a Hickory withe. Burning for hours, such torches lighted the pioneer for miles through the forest at night."[270] Pitch pines are quite pyrogenic, carrying the fire that ensures their regeneration and perpetuation in the landscape.

White pine (*Pinus strobus*), a charismatic species of northern forests, colonizes sunny (often disturbed) ground. This is a rapidly growing species. Easily exceeding 100 feet in height, it grows high above the forest canopy as a mature tree. Its habitat tolerances range from hummocks over swampy ground to steep slopes and summits.

According to the USDA Fire Effects Information System, "[e]astern white pine is moderately fire resistant. Mature trees survive most surface fires because they have thick bark, branch-free boles, and a moderately deep rooting habit. Younger trees are not as fire resistant. The needles have relatively low resin content so are not highly flammable."[271]

Shortleaf pine (*P. echinata*) is a versatile species found in all of the states of the Atlantic Coastal Plain, and also in the Appalachian Highlands and points westward. **Loblolly pine** (*P. taeda*) is a primarily southern species (ranging from New Jersey southward) that is frequently grown in monoculture timber plantations

Pitch pine (*Pinus rigida*).

as southern yellow pine. It is capable of attaining great heights (up to 169 feet tall at the spectacular Congaree National Park). Loblolly is less fire resistant than some of its congeners.

Virginia pine (*Pinus virginiana*) is scrubby, short, and tolerant of harsh conditions; vulnerable to fire but a rapid colonizer of burned landscapes. Virginia pine is present on many of the high bluffs over the Delaware River near me, but many individuals are old and senescing, and are not recruiting in the shade of the other canopy trees. Perhaps fire suppression is at play here, and deer browse of seedlings definitely is. Pine snags provide good, soft wood for excavation by woodpeckers, which in turn can become nesting and shelter sites for many wildlife species from chickadees to tree frogs.

Podophyllum peltatum • Mayapple
Tropical-like fruit, toxic root.

Mayapple is a colonial herb of moist forests, with unusual lobed leaves that sit atop a short stem one or two feet off the ground, looking somewhat like a small parasol or upright propeller. Flowering individuals have bifurcated stems. Mayapple's diaphanous white flower arises from the node where the stem splits. Fully ripe fruits are edible, while the rest of the plant is toxic.[272] The ripe fruit is sweet and has a tropical flavor. It can be hard to come by, as pollination is inconsistent and the fruit is sought after by wildlife. Mayapples located near other, nectar-rich flower species which bloom at the same time as mayapples may benefit from increased pollination due to the attractiveness of the neighboring plants.

Traditional and contemporary herbal uses stress the powerful toxicity of this plant. Podophyllotoxin, found in mayapple, is used in manufacturing etoposide, which is the active ingredient in a drug used for treating lung and testicular cancer. Presently, it is harvested from an Asian mayapple species, but as that species is harvested to near extinction, attention

Mayapple (*Podophyllum peltatum*).

is turning to the American species as a source instead.[273] Let's hope its harvest will be more sustainable. Mayapple saw usage among Indigenous peoples in the treatment of skin cancer and warts.[274] It has traditional uses among the Cherokee as a laxative and cathartic.[275]

Polygonatum: **The Solomon's Seals**

Arching herbs with crisp sweet shoots and starchy rhizomes, for food and medicine.

Solomon's seals are delicious shoot vegetables with a nutritive as well as medicinal rhizome. They have arching stems with flowers in the leaf axils. In the spring, the young shoots arising from the perennial rhizome make a delicious vegetable. Harvest before the clasped leaves at the top of the shoot unfurl, and prepare like asparagus.

The root is sweet and starchy and has some culinary potential. Tinctured, it makes a joint remedy, its use indicated by creaking or fragile-feeling joints. I use Solomon's seal as a component of a back pain formula described by herbalist Jim McDonald and it has worked remarkably well for me. He describes Solomon's seal as, "without a doubt among the most valuable herbs for addressing joint injuries of all kinds." [276] His formula also includes mullein, horsetail, black cohosh, and goldenseal.

Downy Solomon's seal (*Polygonatum pubescens*) is a common herb of mesic and dry forests. **Smooth Solomon's seal** (*P. biflorum*) differs from its relative primarily in the lack of pubescence on the leaves and in its non-clasping. It has significant overlap with downy Solomon's seal in habitat, but smooth Solomon's seal is more often found in the higher-elevation, more xeric habitats in my field experience.

Giant Solomon's seal (*Polygonatum biflorum* var. *commutatum*) manages to look husky and graceful all at once. Spreading by rhizomes, this herb can form a large patch over time. It is particularly large and prolific and ideal for cultivation as a shoot vegetable. In its native habitat it is primarily found in rich, riparian areas.

Pontederia cordata • **Pickerelweed**

An aquatic beauty.

Pickerelweed lofts abundant spikes of purple-blue flowers over shallow waters. Its leathery, beautifully veined arcuate leaves are heart-shaped but more narrow.

Its seeds, which feed waterfowl and muskrats, are also edible for people. They can be boiled, roasted, or dried and processed for flour. Porcher and Rayner write: "The seeds of pickerelweed are a pleasant and hearty food." Young leaves and leafstalks can be boiled and eaten as well. I found them to be quite tough, so seeking out tender material is probably crucial.

A solitary bee named *Dufourea novae-angliae* is a specialist on pickerelweed, foraging pollen only from the plant's beautiful lavendar flowers. The bee's nests are found along the margins of ponds and streams, in sandy soils.[277]

Populus balsamifera • Balsam Poplar

Soothing tree.

Balsam poplar is a moderate-sized, fast-growing tree of northern floodplains and other riverine habitats. Its fragrant winter buds are imbued with resins, polyphenols, and terpenes, and are collected for medicine. Traditional uses for balsam poplar buds include external uses as wound dressings, for frostbite, sunburn injury, joint pain, eczema, and psoriasis, and to prevent infection of wounds and cuts.[278]

Prunella vulgaris var. lanceolata • American Heal-All

Heal all; 'nuff said.

What a great name for a medicinal plant! As one might guess, this is a highly regarded plant in folk medicine. Folk herbalist Tommie Bass was recorded as saying, "Heal-all is a wonderful plant for coughs and colds and for the nerves. The Indians used to say it was good for just about anything, and they should know." Darryl Patton adds that other traditional uses include treatment of shingles and herpes, as well as bronchitis and chicken pox.[279]

Recent scientific study suggests that Tommie Bass and the Indigenous sources he refers to are correct, and the aptly named heal-all has immune modulatory,

antiviral, antiallergic, anti-inflammatory, antioxidant, antidiabetic,[280] anticancer,[281] anti-neuroinflammatory, and antihepatitis[282] properties, and inhibits HIV-1 infectivity.[283]

Heal-all is a well-known weedy species with both native and European-derived varieties. The native variety's main stem leaves are long, narrow, and wedge-shaped at the base, whereas the European variety's main stem leaves are roundly egg-shaped, including a rounded leaf base.[284] As a small, creeping mint family plant, it can tolerate mowing, grazing, and trampling, and can often be found in yards, pastures, gardens, and road edges.

Prunus: The Plums and Cherries

Flavorful fruit, bark medicine.

Prunus is an exceptionally widespread genus of stone-fruit-bearing woody plants, ranging from the dwarf sand cherry to the thicketing American plum on up to black cherry, a full-sized canopy tree. The fresh fruit are choice in almost all species, and medicinal uses for the bark of several native *Prunus* species are well-attested. These are also exceptional wildlife plants, especially as host plants for moths and butterflies, and as fruit producers for consumption by songbirds.

American plum (*Prunus americana*) forms dense, somewhat thorny thickets in disturbed areas with at least part sun. In New Jersey, this puts it in competition with numerous invasive species (e.g., autumn olive, multiflora rose, etc.) that are not as susceptible to deer browse. Even though it is supposed to be a somewhat common species, it took me years of searching to find the first of the half dozen or so populations that I know of.

The fruit can be abundant, but like many wild fruits, colonies may skip fruiting years if conditions are adverse. American plums are about the size of a large cherry, comparable to beach plum.

Records of the usage of American plum by Indigenous peoples are extensive. Out west, the Lakota month corresponding to August is "red plum moon" as this is the harvest month for wild plums. As early as 1895's *Food Plants of the North American Indians* it was noted that American plums were planted by Indigenous peoples in New England and Canada and that "this culture extended to Mississippi" and included varieties selected by Indigenous peoples. The fruits of American plum are eaten raw, but are also dried and consumed over the winter.[285]

Indigenous peoples also use the bark as a medicinal agent. It is used among the Cherokee to make cough syrup (presumably akin to that produced from black cherry, *Prunus serotina*), and to support kidney and bladder function.[286]

American plum grows rapidly and can spread into a large clonal stand. Give it some space. It can also be transplanted by dividing off young dormant suckers and replanting them elsewhere; at least one ethnographic account (from California) describes American plum propagated from slips and/or seeds.[287]

Chickasaw plum (*Prunus angustifolia*) is an occasional coastal denizen that was likely brought east from an original wild range west of the Mississippi. William Bartram astutely noted, "I never saw it wild in the forests, but always in old deserted Indian plantations. I suppose it to have been brought from the S.W. beyond the Mississippi, by the Chicasaws."[288] It has been extensively naturalized in the South, and in some cases remnant populations can be found in proximity to former Indigenous village sites.[289]

Chickasaw plum is a bountiful producer of fruit. Kelly Kindscher reports making some rough calculations that suggest that *Prunus angustifolia* can yield as much as 3,820 pounds of pitted plums per acre.[290]

Herbalist Tommie Bass used the boiled fruit in a sugar syrup for coughs, asthma, and bronchitis, saying, "There ain't nothing better for the asthma than making a syrup out of the wild plum tree."[291]

Beach plum (*Prunus maritima*) is a short, orderly, multistemmed shrub that stays ahead of burial by sands on the back dunes by continually growing and radiating. It can sometimes appear to be a very low shrub, but recall that part of its stems may be under the sand. Beach plum bears clusters of pendant fruit in shades of red, purple, and blue. They look like cheerful little balloons, spherical with a little peduncle radiating from the center.

The abundant little fruits are delicious — tangy, sour, and chewy on the outside, soft and sweet on the inside, with a small stone that releases easily from the fruit. A juice made from beach plums is a warm orange color and absolutely fantastic in flavor.

Given how abundant and delicious the fruit are, how attractive the plant is in flower, fruit, form, and fall foliage, and what tough, resilient individuals are found, it's a wonder that everyone with shore property in the beach plum's native range doesn't landscape with them.

Sand cherry (*Prunus pumila*): This little known, naturally dwarf cherry recommends itself as both a low ornamental (covered in cherry blossoms in season), as well as a delicious edible. Its cherries are sweet and tasty and are the largest of our native cherry fruits. Its fruits are consumed fresh or dried. A western subspecies is used as a component of sauces among the Dakota, Omaha, Pawnee, and Ponca.[292]

Sand cherry spreads on rocky ledges, sandy dunes, and riverine scour via rhizomes, hugging the ground in these areas free of taller, shading plants. Three varieties of the species are recognized with differing foliage characteristics and form, including the prostrate form var. *depressa*, and the somewhat more upright var.

susquehanae. The latter is most likely to be found in glade habitats, and is classified as its own species, *Prunus susquehanae*, in recent manuals.

A third variety, var. *pumila*, is restricted to the Great Lakes region.

Black cherry (*Prunus serotina*) is an edge and gap colonizer that persists and thrives as a canopy tree in some areas. Its fruit are flavorful, rich and sweetish with an almond, bitter overtone. They are also fantastic sun-dried.

The bark is used as a flavoring (think black cherry soda) as well as a medicinal herb indicated for spasmodic coughing (think cough drops). Black cherry is also a topical anti-inflammatory and can reduce the duration of cold sores and other skin irritations.[293] I enjoy making black cherry bark syrup and mixing it with seltzer to make homemade black cherry soda. By the way, the key to making soda that tastes like store soda is adding incredible amounts of sugar. Keeping the sugar level relatively low still makes a fine beverage but it will not have that sweet, syrupy thickness many of us are accustomed to. Tommie Bass's recipe for cherry bark syrup is to "take the inner bark and [low] boil it in water for about thirty minutes or so and then add sugar and boil until it is good and thick." His student Darryl Patton elaborates, suggesting a double handful of bark to one gallon of water to four cups of sugar, and recollects that Bass "believed that cherry bark was good for just about anything that ailed man or beast," including coughs, colds, asthma, bronchitis, nervine, cancer, and tonic in his list of traditional uses.[294]

This is a very versatile species that will colonize edges and open understories as well as act as a canopy tree in some — primarily northern — forests. It has a vast range stretching from Nova Scotia south to Guatemala.

There is a whole ecosystem at work in a mature black cherry tree. Nectaries on its leaf stalks feed ants, which in turn defend black cherry from defoliating caterpillars and other leaf feeders. Eastern tent caterpillars, which are frequently found feeding on black cherry, ingest toxic glycosides from its leaves and defend themselves from resident ants by regurgitating toxic fluids.[295] All in all, black cherry hosts over 450 species of butterfly and moth larvae.[296] This in turn makes it a superb species for supporting native bird populations, all of which feed invertebrates to their young, and some of which eagerly consume and disperse black cherry fruits as well.

Chokecherry (*Prunus virginiana*) ranges in form from a low shrub on shallow soils to a small tree in more fertile ground. It is found in glades and also along the edges of rich woods. It is more common northward and at higher elevations, and I've seen it as a common roadside species in rural northern Pennsylvania and Maine.

Chokecherry
(*Prunus virginiana*).

It handles a variety of conditions, including shade, with flexibility and ranges across much of Canada, and the entire United States barring the Deep South.

Because of its wide range and utility, chokecherry is widely used in Indigenous foods and medicinal practices. Fruits are ground and mixed with fat to make pemmican, consumed fresh or dried, made into preserves, added to soups and sauces, and made into beverages, syrups, and wines. The juice of chokecherries which resulted from processing the fruit for pemmican making was "given as a special drink to husbands or the favorite child" among the Blackfoot.[297]

In his book *The Sioux Chef's Indigenous Kitchen*, chef Sean Sherman includes a recipe for *wojape*, a traditional sauce made by cooking six cups of fresh chokecherries with one to 1.5 cups of water and honey or maple syrup to taste. Sherman describes its use for a "dessert or ... tangy sauce for meat and game and vegetables, and as a dressing."[298]

True to their name, chokecherries can be astringent, especially when not fully ripe. This becomes less of an issue as the fruit reaches full ripeness, and as one develops a taste or tolerance for the mild levels of astringency found in a number of wild fruits.

Like black cherry, chokecherry bark is used as a respiratory relaxant and antispasmodic. Herbalist Jim McDonald describes it as "a cooling sedative to lung tissue, [which] excels when heat and irritability undermine healthy expectoration."[299] It is used in conditions of heat, excitation, and irritability, and should not be used in overly relaxed tissue states. Herbalist Kiva Rose recommends its use as a tonic for the heart, including anxiety and palpitations.[300]

Chokecherry can expand to form thickets. It is drought tolerant, and somewhat salt tolerant as well, making this a good species for planting on high road banks.

Pseudognaphalium obtusifolium • Rabbit Tobacco

Smells like green spirit (and a bit like IHOP). This inconspicuous annual has distinctive silver-haired stems. After handling, it leaves an aroma on your hands that reminds me of (artificially flavored) pancake syrup. It is a highly regarded medicinal herb in Southern Appalachian medicine, a terpene-rich herb used in smoking mixtures and teas to dry the lungs and bring up phlegm. It is used in cases of asthma, bronchitis, and pneumonia, and was highly regarded by renowned Southern folk herbalist Tommie Bass.[301] It may seem odd to smoke a medicinal herb for health benefits, but recall that air brought into the lungs only needs to pass though a barrier two cells thick to enter the bloodstream and from there the body. In the case of rabbit tobacco, the benefit is even more immediate, as most of its uses pertain to the bronchial tubes.

According to herbalist David Winston, rabbit tobacco is known to the Cherokee as *katsuta equa*, and frequently used as a remedy for coughs, colds, flu, strep throat, and fever in children.[302] Herbalist Matthew Wood utilizes the herb for "lifelong and congenital asthmatics."[303]

Pycnanthemum: The Mountain Mints
Frosty wild mints.

Hand a stem of one of our *Pycnanthemum* species to a complete botanical beginner and they will have little trouble guessing that it is a mint. Not because of its square stem, opposite leaves, or irregular flowers, all typical of the Lamiaceae, but because of the aromatic volatile oils that instantly identify many mints and make mint species culinary, tea, and fragrance mainstays.

The essential oils in mountain mint species, primarily menthol, menthone, and limonene, have a cooling minty aroma. Many of the important constituents in mints are contact antimicrobials.[304] They act as stimulating movers in the human body and are employed to reduce stomach cramping.[305] The leaves of the plant make excellent beverages (tea, iced tea, mint julep). The leaves can be used in a steam and inhaled to relieve congestion. Taken internally, mountain mints act as expectorants and sedate the bronchial nerves. The Cherokee use a poultice of hoary mountain mint for headaches, and a leaf tea for fever, and the Lakota use Virginia mountain mint tea for coughs. The tea is diaphoretic and carminative.[306] Mountain mint species can also lend their strong aromatics to mosquito repellants.

With small, easily accessible flowers, mountain mints support numerous pollinators including butterflies such as skippers, hairstreaks, blues, fritillaries, as well as native bees, bumblebees, wasps, and flies. Mountain mint flowers are clustered in tiers that bloom sequentially over a long period of time. They are considered to be one of the most pollinator-friendly plants and are often planted to support native bee populations.

Virginia mountain mint (*Pycnanthemum virginianum*) is found in dry to moist meadows and openings.

Narrowleaf mountain mint (*Pycnanthemum tenuifolium*) is found in similar habitats to Virginia mountain mint. They are similar in form but narrowleaf mountain mint is shorter, with more slender leaves. Narrowleaf mountain mint is considered by the Xerces Society to be one of the best wildflowers to support native bee, beetle, and butterfly species.

Broadleaf mountain mint (*Pycnanthemum muticum*) is fairly shade tolerant and can be found on edges, in young woodlands, and in small openings, usually on moist soils.

Hoary mountain mint (*Pycnanthemum incanum*) is a tough species native to open rocky slopes and glades, where it blooms in

Hoary mountain mint (*Pycnanthemum incanum*).

in the woods. Often, mature oaks are found surrounded by younger forests, legacies from the days when the oak stood in an open pasture, providing shade and acorns for livestock.

Acorns are a prolific source of high-quality starch, protein, fats, and certain nutrients. A food system based around acorns would likely be more ecologically sustainable than one based around annual grains. Creating culture around oaks and the food they give us is as practical and important a sustainability path as any. However, as inheritors of a culture more intent on cutting trees than tending them, we have some serious systemic inertia to overcome.

Human cultures across the world have subsisted through balynophagy, the consumption of acorns as a staple crop. In North America, this lifeway has frequently included not just acorn consumption but Indigenous peoples' active manipulation of habitats such that they support open, ancient groves of highly productive oaks. While the resulting structure may resemble an orchard in some ways, the groves are not necessarily created by planting but through the management of the underlying conditions that produce and maintain oak ecosystems.

The oak stewardship lifeway extended across the continent, from open woods in pre-colonial New England that you could ride a horse through, to the live oak groves of Florida and California. The primary

tandem with associates such as woodland sunflower and upland boneset to create a pollinator's paradise and a glorious show.

Quercus: The Oaks

Staple food producers, majestic trees.

Oaks are among our most distinctive forest species. An elder oak always draws my attention even after many years of working tool of the oak-stewarding cultures is fire, which abets oaks in a variety of ways. By thinning competing and younger trees, fire allows for open, spreading oaks, which are prolific and easy to harvest from. Fire also consumes litter underneath fruiting oaks, destroying acorn-consuming weevils and infested fruit as well as other pests and pathogens that afflict oaks or decrease the quality of the acorns they produce. In the absence of fire stewardship, oak habitats decline in health and oaks are often replaced by mesophytic, shade-tolerant species such as beech and sugar maple.

All oaks produce acorns, and all acorns have tannins that deter their consumption. Despite these astringent compounds, more than 180 species of birds and mammals consume acorn meat, ranging from red-bellied woodpeckers and wild turkey to bears and raccoons. Humans have developed methods for processing acorns that leach out the water-soluble tannins and yield a versatile acorn meal or flour.

We use a very simple method to process acorns for flour and meal, which we learned from Julie Martin and Eddie Starnater. In brief, it looks like this:

1. Shell acorns.
2. Grind acorns to coarse meal.
3. Fill large jar(s) ⅓ with acorn meal.
4. Top off jar with cold water.
5. Agitate jar.
6. After the meal settles and the water is no longer cloudy, pour off water. It will look dark, especially the first few times. Usually it takes a few hours for the meal to settle.
7. Repeat steps 5–7 until the water no longer darkens, and the meal doesn't taste bitter. Usually five or so repetitions are all that is required.
8. Dry the meal, and grind it finely if flour is desired. Store in airtight containers.

This is essentially a process of repeatedly soaking and decanting — about as simple as it gets. It would not be a challenge in our industrial society to develop simple machines to crank out acorn meal by the ton if this were a cultural priority.

The advantages of consuming acorns as a staple crop are numerous compared to starchy annual crops such as wheat, corn, rice, and potatoes. They involve the stewardship of an intact, diverse oak ecosystem rather than the management of monoculture fields.

Oaks also have medicinal uses, usually attributed to tannins, which bind to proteins and can have antibacterial properties. Extracts from white oak galls inhibited the growth of drug resistant *K. pneumoniae* bacteria in one study. Other constituents in white oak may be responsible for inhibiting quorum sensing in drug-resistant bacteria.[307]

Herbalists use the bark of oaks, especially white oak, as medicinal agents based around the astringent properties of the bark. White oak bark tea is used for diarrhea, bleeding, sore throat,[308] and a host of other conditions where an atonic, overly relaxed tissue state of the body needs tightening, toning, and drying.[309]

Some of the most widespread and large-fruited oak species are summarized below.

Quercus alba, **the white oak**, can live up to 600 years and produce 10,000 acorns in a mast year.[310] Its bark is relatively light in color and, on a foggy winter's day, it looks like a silver ghost.

Oak leaves decay slowly, are low in nutrients,[311] and can suppress ground layer herbaceous plants. Part of maintaining diversity in oak-dominated habitats can include controlled burning. The bark of white oak shields it well against fire damage.

Quercus bicolor, **the swamp white oak**, stands out even in a genus full of beautiful trees. Its majestic trunks have exfoliating bark (peeling sideways in long sheets), and its large acorns have a reputation for fairly low tannin content. It is able to tolerate periodically inundated habitats such as those found along creeks and rivers and in forested lowlands. *Quercus michauxii*, **swamp chestnut oak**, is very similar but has a more southerly range.

A colleague with extensive experience in wetlands restoration reports that swamp white oak is one of the most reliable tree species for survivorship in restoration plantings.

Quercus lyrata, **the overcup oak**, is found in lowlands in the Coastal Plain.

White oak (*Quercus alba*).

Its acorns are fairly large, growing one to two inches in length. The acorn cap covers more than two-thirds of the acorn, rendering it more buoyant for seed dispersal.

Quercus macrocarpa, **the bur oak**, has exceptionally large acorns. Generally a midwestern species, outliers (and remnants of managed groves)[312] exist in the East. In New England, it is found in moist bottomland conditions as well as xeric upland sites. It prefers calcareous (limestone) soils. In the Midwest, it is associated with oak savannas, fire-dependent ecosystems that had nearly vanished before their rediscovery in the late 20th century. It has the thickest bark and is most fire-tolerant of the oak species. In parts of Texas, bur oak can produce large acorns the size of golf balls.

Quercus montana, **the chestnut oak**, is an oak of steep, rocky slopes, with craggy, deeply furrowed gray bark that looks contiguous with the bouldery breakup below it. This species stands out for its large acorns. Like other white oak group species, it sends out a young root radicle shortly after falling to the ground, necessitating immediate processing (or freezing) if using acorns for food.

Chestnut oak's thick bark helps to protect it against fire. Its drought tolerance makes it a viable species on all but the most thin-soiled, xeric sites.

Quercus rubra, **the northern red oak**, is a widely distributed species found in mesic and loamy sites. As is characteristic of the red oak species group, it takes two years to develop each acorn crop. The resulting acorns are fairly large, and remain dormant until the spring, unlike white oaks. Northern red oaks begin to fruit at 25 years old, though sizable crops usually occur on species over 50 years in age.[313]

Rhodiola rosea • Roseroot

Boreal medicine.

There is a spectacular trail in Downeast Maine where the thick conifers and deep mosses of the boreal forest abruptly give way to stark seacoast cliffs. The Atlantic Ocean thunders below, carving abrupt canyons in the rock faces. Here, clinging to high crevices in the rock, bathed in near-constant cooling mists, resides roseroot. It stands out immediately, with pink-red flowers atop blue-green succulent leaves and stems. This Arctic species is found in a few northern states, its primary habitat being cool, moist cracks in cliffs and on rocky ledges. Remnant populations in Pennsylvania, and in one county in North Carolina, probably date back to the Pleistocene when colder habitats prevailed throughout the region.

Roseroot is a circumboreal species found in Europe and Asia as well. Recorded use in Tibet and China dates back at least a thousand years. It is an adaptogen used to help the body cope with stress and improve performance; in Russia it was used by Olympic athletes and by cosmonauts to boost physical and mental acuity. It is a sweet and slightly bitter herb frequently utilized in tea or tincture form.[314]

Because it is revered as a medicinal plant, it is threatened by overharvest despite its wide distribution. Progressive herbal distributors are seeking cultivated American plant material to reduce the pressure on wild populations, which are subject to legal protection in a number of states and countries.[315]

Rhus glabra • Smooth Sumac
Rhus typhina • Staghorn Sumac
A lemonade stand in a disturbed habitat.

I always know when the road edges in Pennsylvania have been sprayed with herbicides because the sumac foliage starts turning brilliant orange and scarlet. Normally these are sumac's fall colors, but this usually happens sometime in midsummer. For some reason this plant draws the endless ire of road crews. It's often lumped in with the invasive *Ailanthus* as a "weed tree" or "railroad tree." These pejoratives are somewhat apt, as sumac thrives on cinders and other inhospitable, gravelly, dry ground. It is not quite a tree however, but a multi-stemmed shrub that can occasionally reach 15 or more feet tall.

The distinctive upright clusters of sumac's red fruit are a familiar sight, and many people are aware that they can be brewed into a "lemonade" of sorts — tangy and high in vitamin C. I've ground the lemony fruits to make a spice resembling Middle Eastern *za'atar*, a tart seasoning derived from an Old World *Rhus* species. If you rub your hand along the ripe fruit cluster of a staghorn sumac, you'll get a sneak preview of this wonderfully citrusy spice to lick off your fingers.

Forager Sam Thayer describes peeling and eating sumac shoots, which can be quite prolific; the plant is a superlative spreader and resprouts effusively after cutting. The first few times I tried it I enjoyed the crunchy, somewhat sweet flavor; more recently, I've found bitter flavors to be deterrent. Likely the shoots need to be fairly young or they will have developed tannic flavors. Consumption of the young shoots of smooth sumac is documented for the Iroquois in New York. Peeled smooth sumac stems can be used as chew sticks, and contain an antibiotic phytochemical that can prevent tooth decay.[316]

Anishinaabe plantswoman Mary Siisip Geniusz describes a potent yellow dye that can be derived from the pith of the sumac's stems. The stems are easy to hollow and we've used them as DIY maple sugaring spiles.

Herbal uses of sumac are diverse and utilize the seedheads, inner bark, and sometimes the leaves. Herbalist Matthew Wood lists a number of indications for sumac, including profuse sweating and urination, diarrhea, chronic bladder catarrh, and swollen prostate. He describes it as "the superlative remedy for stopping excessive flux from any channel of elimination."[317] Darryl Patton recommends the use of sumac bark in the treatment of influenza.[318]

Sumacs excel at restoring highly degraded sites where more conservative species would struggle. Restorationist Margaret Gargiullo suggests these species as "primary or secondary species for initial vegetation of eroded slopes, fill, landfills, roadside banks." Once established, they can act as "nurse" plants for tree species to recruit or be added later, which will eventually shade out the sumacs.

Ribes: **Gooseberries and Currants**

The maligned fruits.

The tragedy of the *Ribes.* Our native currants and gooseberries were subject to eradication programs in the early 20th century because some *Ribes* species are alternate hosts for white pine blister rust. At one point, up to 11,000 Civilian Conservation Corps men were employed in a *single year* to eradicate *Ribes* from the wild.[319] As a result, many of these superb native berry species are rare or endangered in the wild.

In general, both gooseberries and currants seem underappreciated in the United States. At my grandparent's house in Hungary, the fencerow was lined with red currants and gooseberries, deliciously sour and sweet. I have fond memories of walking along the fence, gorging on the ruby-like *ribizli* and grape-colored *pöszméte.* Perhaps the onus of the white pine blister rust has cast a permanent shadow over these species. However, our native flora contains a good number of excellent-tasting gooseberries and currants. The former tend to be prickly-stemmed, whereas many of the currants are unarmed.

If you're interested in restoring habitats using native edible species, the *Ribes* are a great genus to know.

Native Gooseberries

Our native gooseberry species are shade-tolerant, short shrubs, lightly to heavily armed with prickles. In the case of **prickly gooseberry** (*Ribes cynosbati*), even the fruits are prickly (especially when unripe), calling to mind nasty medieval weaponry. However, the taste of these fruits is superb — tart and sweet. These species have high wildlife value as well, with fruit eaten by a variety of birds and mammals. The gooseberries are not known to be carriers of white pine blister rust and are permitted to be planted in some states where the related black currant is restricted, including my home state of New Jersey.

I first met **American gooseberry** (*Ribes hirtellum*) scrambling across rock ledges on the sea-sculpted cliffs of Downeast Maine. Its habitats are variously described as "rocky woods and shores,"[320] and "bogs, fens, forests, meadows and fields, shores of rivers or lakes, swamps, [and] wetland margins,"[321] suggesting another tough customer in the genus *Ribes,* one that thrives best in habitats where competition is minimal due to the adversity of conditions. The fruits are tasty, sweet with a bit of tanginess and fairly typical of the wild gooseberries, which have a generally similar (and always enjoyable) flavor to me.

Appalachian gooseberry (*Ribes rotundifolium*), is a fairly common species that I first encountered in the cleft of an ancient sugar maple tree in the Catskills,

American black currant (*Ribes americanum*).

once a pasture tree and now a wolf tree within a younger forest. Also resident in the tree was a family of great horned owls, and following them led me to the gooseberry, about ten feet off the ground. This is a versatile species that prefers rich, rocky woodland habitats.

I encountered **Missouri gooseberry** (*Ribes missouriense*) for the first time while doing rare plant surveys on behalf of a local conservation group fighting a gas pipeline. The home range for this species is predominantly west of the Mississippi. It seems highly likely that eastern populations were brought here by Indigenous peoples. The site where I first identified this species had been a significant Lenape village site in pre-colonial times. The species is quite similar to Appalachian gooseberry in structure and fruit. If planting it, be mindful that the currently known eastern sites may represent genetically and culturally important populations, and that any introductions could influence the genetics of these isolated, edge-of-range populations.

Native Currants

American black currant (*Ribes americanum*) is a low (waist to head-high), thornless berry bush with tasty, tangy fruit good for beverages, preserves, and raw eating. It is fairly shade tolerant and thrives in moist conditions such as swamp edges, springs, seeps, and even wet ditches. If you live outside of the primary natural range of white pine, it is an excellent choice for a fruitbearing understory shrub.

It seems black currants are susceptible to the blister rust pathogen, *Cronartium ribicola*, though it is the European black currant that is most so, with American black currant (*Ribes americanum*) having some individuals that "show no signs of the disease [and] … generally less susceptible than are European black currants,"[322] according to horticulturalist Lee Reich. Some states ban the planting of some or all *Ribes* species, and in Maine, where the white pine timber crop is considerable, active eradication of wild *Ribes* species is encouraged.

Other currants found in northerly and high-elevation forests include bristly black currant, skunk currant, and wild red currant. For a variety of reasons described below, these might not be one's first choice for edibility or planting.

Naturalist Michael Kudish describes **bristly black currant** (*Ribes lacustre*) as "scarce because of eradication," and native to "well-drained sites, shade-tolerant, under northern hardwoods usually, but occasionally in spruce-fir" in the Adirondacks,[323] while its Pennsylvania habitat is described as cool, wet woods and swamps.[324] In my experience, the fruits are not tasty. **Skunk currant** (*Ribes glandulosum*) is aptly named: the fruits smell and taste skunky, in my limited experience, though Arthur Haines comments that they are "quite tasty when ripe"[325] so

maybe the fruits I tasted were somewhat old. It is a species of edges, swamps, and ledges in cold climate zones. **Wild red currant** (*Ribes triste*) is a carrier of white pine blister rust. It is listed as threatened or endangered in at least six states. Any restoration efforts should take both these factors under serious advisement.

Robinia pseudoacacia • Black Locust
A bee tree, makes great fenceposts too.

I've always wondered how a tree so dense could grow so fast. Black locust is over twice as durable as white oak, and a cord of firewood has almost the equivalent fuel value of a ton of anthracite coal. Perhaps it is because of this tree's ability to fix atmospheric nitrogen in partnership with bacterial symbionts.

Black locust is a pioneer tree in openings and disturbances on mesic slopes. Its original native range appears to have been "cove and mixed mesophytic forests in the central and southern Appalachian Region."[326] It has been widely dispersed and planted by people in the past few centuries, and black locust is now widely naturalized throughout the Northeast. It is also invasive in parts of Europe, Asia, South America, Africa, and Australia.[327] As a kid visiting my grandparents in Hungary, I relished the clear honey known as *akácméz* — acacia honey, the "acacia" in question being black locust. The tree comprises something like 20% of the entire forest cover in Hungary

Black locust (*Robinia pseudoacacia*).

America where it can quickly dominate open, full-sun plant assemblages.

Let's talk more about this remarkable tree's virtues. Its wood is highly decay resistant, useful for in-ground posts as well as durable tool parts. It also produces delicious flowers, which are sweet like peas and light like flower petals. Black locust blooms abundantly, with large, showy white flowers, and gathering a bowlful for a salad, pasta garnish, or snack is easy, provided low branches to harvest from.

Black locust generally matures to a medium-size tree (40–60 feet) and is fairly short-lived — under a century in most cases. Its peak nitrogen fixation may occur fairly early in its life cycle (one study documented that it peaked at 66 lbs/acre/year at 17 years of age).[329] It is not particularly shade tolerant and dies out of maturing forests under the canopy of larger, longer-lived species.

Rubus: Raspberries, Blackberries, and Dewberries

The transitional fruits.

I think of our native *Rubus* species as embodying transition. They have a lifestyle somewhere between herb and shrub, with woody canes that live two years and then die back. These canes flower and fruit in their second year, and often arc back towards the ground to root at the tip. *Rubus* are colonial species that travels across the landscape in a slow wave. *Rubus* often

and is widely used for fuelwood and timber.[328] One should exercise caution about introducing this species outside of its native range, especially into or in proximity to high-quality native habitats. It has been found to be especially problematic in barrens and prairie-type ecosystems in North marks the transition from herbaceous dominated young habitats to older habitats dominated by woody plants.

Blackcap raspberry (*Rubus occidentalis*) fruit resemble red raspberries, but are nearly black when ripe, and are smaller and firmer than domesticates. The canes are distinguished by glaucous stems which appear almost purple in hue. Blackcap raspberry is commonly found on edges and reverting fields throughout our region.

Growing at the edges of woods and fields in cooler, northern and mountain habitats, **red raspberry** (*R. idaeus* var. *strigosus*) is essentially the same plant as those that bear familiar, store-bought raspberries. Somewhat shade tolerant, the plant will bear best in a sunny area.

In addition to delicious fruit, wild raspberry leaves make an exceptional wild tea. Its nutritive and tonic properties cause it to be recommended for pregnant women[330] but it is worth enjoyment by all. Because of its mild tannins, it is one of our better black tea substitutes, and is also a remedy for mild diarrhea, bleeding gums, and mouth ulcers.[331]

Allegheny blackberry (*Rubus alleghe-niensis*) has tall, arcing stems armed with recurved prickles. The stem appears fluted. Its leaves are divided into five leaflets and are palmate. I find its fruit to be very variable. Sometimes it is sour, dry, and coarse, but sometimes the flavor is wonderfully sweet and rich. The best wild blackberries

I've ever tasted were on an exposed summit in full sun in 95 degree weather, during a drought. The sheer volume of heat and sun that year created truly delicious fruit. I don't know whether insipid fruit is the result of shading, weather, or genetic variability, but a combination of the three seems logical. Botanists recognize many blackberry species, but *Rubus alleghenien-sis* is by far and away the most common species I encounter while doing fieldwork. Allegheny blackberry's stout canes and sharp, recurved prickles make excellent shelter for small mammals and birds. It can also be used as a natural deterrent to human access to certain areas that are to remain protected from foot traffic; the Central Park Conservancy in New York had us custom grow this plant for several years for this exact purpose.

The **dewberries** (*R. hispidus* and *R. flagellaris*) are low, sprawling groundcovers spreading on decumbent stems. Usually they are a decent nibble but not exceptional. But I've tasted big delicious dewberries on occasion, and as a fruiting groundcover they are worth consideration. Michael Kudish, in the Adirondacks, says, "the fruits [of *R. hispidus*] are small but sweeter than those of blackberry and raspberry, and well worth the effort of harvesting."[332] Note that *Rubus flagellaris* can be stout and prickly and a field full of it can offer a mild challenge to walkers, at least in sandals and shorts.

Purple-flowering raspberry (*Rubus odoratus*) is a big beauty of a shrub, with wide soft palmate leaves and large purple flowers that somewhat resemble those of wild roses. Unlike the *Rubus* species above, purple-flowering raspberry is thornless. This species often grows along edges and gaps in mountain forests, such as along the high banks of streams, hanging off of shaded bluffs, or along the edges of woods roads in dark hemlock forests.

Its fruits are sweet and flavorful, with a dry, seedy crunch. They are better off taken on their own merit rather than compared to the familiar domestic and wild raspberry *Rubus idaeus*. I greatly look forward to their crunchy tang every year.

Rudbeckia laciniata • Tall Coneflower

Sunny wildflower, edible greens a spring treat.

This is a tall and beautiful wildflower somewhat reminiscent of prairie perennials such as *Echinacea* and black-eyed Susans. Its core habitat is in sunny breaks and meadows along watercourses, where it can grow to impressive heights in the company of other flamboyant forbs such as Joe Pye weed. Its flowers are comprised of cone-shaped clusters of disk florets with showy reflexed yellow rays.

Tall coneflower's tender new leaves are edible; choice harvest seasons include early spring as well as fall, when basal leaves put on flushes of growth. These greens are plentiful during times when other greens are scarce. The leaves are best prepared by cooking with bacon grease or lard, but frying can be done in any oil. The cooked herb is known as sochan among the Cherokee and is an important aspect of a traditional spring diet, coming out of wintertime when plant foods are much less available and diverse.[333]

The seeds of tall coneflower are highly sought after by goldfinches and other seed-eating birds.

Sagittaria latifolia • Broadleaf Arrowhead

Duck potato.

From shallow, slow waters emerge broad, upright leaves, pointed like arrows with rear lobes thrust backward and towards the marsh or pond surface below. In the midst of these sagittate leaves arises the flowering stem, bearing a raceme of open, white, three-petaled blooms around a ring of golden anthers. Deep beneath the water lie rhizomes that enable colonies of this plant to form, and along these lie tubers that form the prime edible portion of the plant.

Also known as duck potato and wapato, these ovoid tubers are a delicacy, edible like other starchy tubers (think potatoes) but with a sweet taste all their own. Other parts of the plant, including young foliage and young shoots are also edible. Forager Sam Thayer calls the cooked young greens "excellent."[334]

Salix nigra • Black Willow

Craggy willow relieves pain.

This is a fast-growing willow tree, charmingly eccentric in appearance, shaped by the vagaries of flooding and shade.

Black willow is used by herbalists as an anti-inflammatory; it contains salicylic acid, from which aspirin is derived. Its traditional uses include headaches, arthritis, fevers, and other inflammatory conditions.[335] Its light, flexible wood is used in basketry and wickerwork.[336]

Though many willow species have interchangeable uses, Indigenous peoples select different species for their particular utility. For example, among the Arikara, *Salix amygdaloides* was used for basketry; *Salix eriocephala* for cooking wood; and *Salix interior* for fish traps. Willows in general were used to make the roofs of earth lodges, together with bundles of big bluestem grass.[337]

The easiest mode of reproducing black willow is to simply stick cuttings deep in the soil. Willow will root easily, an adaptation to being a marsh and floodplain species; broken branches can be carried to new locations by flood waters and root in to create new stands of willows. Willow cuttings can be used as live stakes, bundles and fascines for stabilization of stream banks. It has dense root systems that are ideal for erosion control. Auxins (plant hormones) created by willows encourage rooting even in non-willow species, and some people cold brew a "tea" of willow cuttings and then soak cuttings of other species in the liquid as a natural rooting stimulant.

Willow flowers are among the very first blooms of spring, providing an important source of pollen for native bees. Willows also host numerous butterfly and moth species.

Sambucus canadensis • Black Elderberry

Revered healer, grows in almost any wet spot.

Found along sunny watercourses and shores and in marshy meadows, black elderberry is a fast-growing shrub that produces prolific white platter-like flowers followed by deep purple-black fruit.

Black elderberry (*Sambucus canadensis*).

Both the flowers and fruit are used in food and medicine; the former as a tea, syrup, or wine ingredient, that latter in syrups and jellies. The flowers have a pleasant floral sweetness. They are used by herbalists as fever reducers suitable for use with infants.[338] The cooked fruit (and other plant parts, seldom used except by advanced practitioners) have a variety of antiviral and anti-inflammatory effects.[339] In Appalachian folk herbalism, elderberry juice is used as a blood purifier, and folk herbalist Tommie Bass used elderberry salve to remedy skin conditions such as eczema and psoriasis. Hot elderflower tea is employed as a diaphoretic, inducing sweating, and a cold infusion is used as a diuretic, promoting detoxification through urination.[340]

Elderberry readily roots from dormant stems. These can be harvested in winter and stuck in the ground during the dormant season and will root while leafing out in spring. I usually harvest stems in February and stick them as soon as the soil is thawed. Using elderberry in this fashion as "live stakes" is ideal for cost-effective stabilization of stream banks and other erosion-prone habitats.

Elderberries fruit on first-year wood. They can be pruned to the base and still harvested from the same year, allowing for easy-to-reach fruit and flowers. Numerous frugivorous birds and mammals feed on its fruits as well.

Sassafras albidum • **Sassafras**
The root beer tree.

Sassafras is a medium-sized, colonial tree that pioneers forest growth in open habitats and is frequently found along sunny edges. A line of living or dead sassafras found in the woods suggests an old hedgerow subsumed by regrowth forest. With its soft wood, dead sassafras trees are easily excavated by woodpeckers, who create excellent habitat for other wildlife in the form of their nest holes. Sassafras fruits (drupes) are highly sought after by birds and are a high-lipid food source for them.

The traditional human usage of sassafras is in tea form. In the Appalachians it is valued as a tonic beverage. Sassafras tea has a delicious flavor and aroma; digging a root out of the ground, it's amazing to smell the essence of root beer! As food, sassafras leaves are dried and powdered to make a stew thickener used in gumbo. Eaten fresh, the mucilaginous (moist and thickening, in this case) quality of the leaves is evident.

One year, I was doing botanical surveys in late winter, monitoring some sites that were due for controlled burning. It was still cold, especially in the morning, and three fieldwork days in a row I felt like I was battling a cold. I poured a liberal amount of sweetened sassafras syrup into a thermos with hot water, and just nursed on that thermos all morning long. It kept me warm, energized, and kept my cold at bay. Herbalist Tommie Bass recommends a hot sassafras tea to help one sweat and "get rid of fevers and colds."[341] Sassafras once enjoyed widespread use in the South as a spring tonic, used to catalyze circulation to the extremities after a winter of slow, sluggish blood stored in the core.[342]

Unfortunately, sassafras is currently prohibited as a food ingredient by the FDA due to the presence of safrole. When dosed to lab rats as an isolated compound, safrole is a low-grade carcinogen. Among herbalists and foragers, the purported risks of sassafras are frequently contested, on the basis of the poor water solubility of safrole. For example, chemist Lisa Ganora writes: "Sassafras tea has not been associated with the development of cancer in humans; it would be difficult to get significant doses of safrole from a water infusion".[343] A long history of unproblematic traditional use supports this claim. That said, at least one other source suggests that even sassafras tea can contain significant sources of safrole.[344] The use of sassafras leaves for gumbo file (thickener) is generally regarded as safe.[345]

Scutellaria lateriflora • Mad Dog Skullcap

Small herb big calming.

Mad dog skullcap is a short herb with small flowers and a very cool name, originating with its former use as a remedy for rabies. That usage is questioned by many, but there may be at least a glimmer of truth to it. Even if it *doesn't* cure rabies, mad dog skullcap's flamboyant name may be a good way to remember its primary herbal usage — as a calming nervine and antispasmodic. Herbalist Matthew Alfs calls it the "remedy *par excellence* for nervousness, restlessness, anxiety, worry, PMS tension, and neurasthenia (nervous exhaustion)." He also recommends it for nervous disorders involving twitching, tremors, and other irregular muscular action.

Given this herb's support for both nervous conditions and spasmodic movement, the connection to rabies has some logic. But more importantly, here is a native herb that offers relief to stress-related nervous conditions, can act as a nervous system restorative, and holds promise for spasmodic and convulsive chronic conditions.[346]

A Chinese relative, Baikal skullcap (*S. baicalensis*), has well-documented antiviral properties.[347] Unfortunately, mad dog skullcap is something of an unknown quantity in terms of potential antiviral activity, with herbalists disagreeing as to its possible utility. For example, herbalist Stephen Harrod Buehner states: "there is reason to believe that the increase in pharmacological action of Chinese skullcap over the usual American species is due to the difference between using the root in Chinese practice and the leaf in American practice." He does not, however, end up recommending the use of mad dog skullcap as an antiviral.[348] The issue seems unresolved pending further research.

Sium suave • **Water Parsnip**
Emergent food.

Perhaps because it is related to (and thus, somewhat resembles) the highly poisonous wetlands species water hemlock, this species is not well-known as a food plant. The confusion might be confounded by its alternate name, hemlock water parsnip (yeeps!), and also by the variability of its foliage: the plant produces differing aquatic, emergent, and terrestrial foliage. Its use by Indigenous peoples, however, was widespread among the Quebec Algonquin in the Northeast and a number of peoples further west. The most common food use of the plant is the use of the root as a raw or cooked vegetable, but some usage of the foliage is also documented. Harvesters should take care to note and avoid the presence of water hemlock when foraging for this plant.

Water parsnip is commonly found growing in the channels of forested headwaters streams, with companions such as cardinal flower, tussock sedge, monkeyflower, and purpleleaf willowherb. It is also found in marshes, ditches, and other sunny to partially shaded wet habitats, and can grow as an aquatic emergent in water up to a depth of one and a half feet.[349]

Smilax herbacea • **Carrion Flower**
Smells like death, tastes like asparagus.

OK, the flower of this plant smells like dead roadkill deer. Its pollinators are flies and other carrion feeders. That said, its flowers bloom on beautiful spherical umbels borne on a gentle herbaceous vine. It lacks the prickles of some of its congeners. Its tip growth and fresh shoots are said to make a fine vegetable. I look forward to trying it, but the deer around here relish it so much that I'll have to germinate and grow my own before I can in good conscience consume it myself.

Carrion flower (*Smilax herbacea*).

Botanist Tim Spira reports that "Native Americans used carrion flower's roots and leaves for medicine, the woody rhizomes for smoking pipes, and the seeds as beads." [350]

Solidago odora • Blue Mountain Tea
The sweetest goldenrod.

Many goldenrods make a decent tea, useful in remedying allergy symptoms like a runny nose and itchy eyes. They have a terpene flavor, decent but unexciting. Blue mountain tea on the other hand has a wonderfully sweet anise aroma and makes a delicious tea. It is a modest-stature goldenrod species that thrives in poor soils where taller competition is absent.

Solidago sempervirens • Seaside Goldenrod
Salt-tolerant goldenrod.

This goldenrod is an important component of seaside plant communities, and occupies the landward side of primary dunes and other dunes to the rear of the seaward dune faces. It blooms late in the fall and provides nectar for migrating monarch and other butterflies. It bears multiple upright flowering stems and large blooms, making it especially attractive. Like other goldenrods, the foliage and flowers can be made into a pleasant tea. Matthew Wood writes that goldenrods are good for treating allergy symptoms, especially those featuring itchy eyes and a runny nose.[351]

Huron Smith documented the following usage for goldenrods among the Flambeau Ojibwe: "Goldenrod leaves and flowers have at times held a rather important place in materia medica, for their carminative, and antispasmodic properties. They have also been used as an intestinal astringent." [352] Likewise, for the related species *Solidago speciosa* among the Meskwaki, Smith writes "[t]he root of this is used to make a tea that heals burns or scalding from steam." [353]

Thuja occidentalis • Northern White Cedar
North Woods cedar.

Kayaking a lake margin in Maine, my eyes are drawn from the exposed rock on the banks to the trunks of trees that seem contiguous with the rocks themselves, with grey, tightly striated bark. Bearing evergreen foliage in flat, ferny sprays, this aromatic tree embodies the North Woods for me.

Further south, northern white cedar can be found as a rare species, often associated with limestone stream edges and north-facing bluffs. Like other northerly species, it is not well-accustomed to the unrelenting heat of a mid-Atlantic summer.

This beautiful tree features prominently in the worldview of northern Indigenous peoples. Mary Siisip Geniusz shares that the Anishinaabe consider that Grandmother Cedar keeps the lines of communication open between humans and the spiritual world. In this regard it is used as an incense, and as an anointing oil, used at birth, death, and times of need, as well as during the sweat lodge. It is often paired with bearberry (*Arctostaphylos uva-ursi*, see above) in kinnikinnick, "a mixture of pleasant-smelling herbs, the most pleasing of the abundant flora, the most pleasing to human and therefore, hopefully, the most pleasing to the Divine." [354]

Northern white cedar (*Thuja occidentalis*).

Cedar is also used to produce cordage, as an emergency food source yielded from the inner bark, for fire-starting drills, and for mold-resistant woven baskets and bags. [355] Cedar yields a healthy tea but some sources recommend moderate usage due to the presence of thujone, a toxic terpene. Cedar's phytochemistry also includes medicinally active glycosides, tannins, and mucilage, which act as "antibacterial, antiviral, antifungal, immune-stimulating, expectorant, and counterirritant" agents according to botanist Arthur Haines. [356]

Tilia americana • Basswood
Gentle giant with broad, edible leaves, flowers a calming medicine.

Basswood is found in the canopy of diverse floodplain forests, and sparsely on mesic slopes and ridgelines. It is a codominant with sugar maple in mesophytic forests of the upper Midwest. [357] Also known as linden, this tree has large, asymmetrically heart-shaped leaves. It is often multi-trunked, an unusual default growth form for a canopy tree. A large individual along the Musconetcong River near me has 13 trunks and its base occupies an area as large as a car, but ancient basswoods like this are seldom found.

One of relatively few insect-pollinated canopy trees (many are wind-pollinated), basswood bears bell-shaped fragrant flowers that are very attractive to bees. The flowers make a nice tea with calming properties.[358] The leaves provide a lesser known but excellent food. The freshly expanded leaves in spring are tender and have an excellent texture and flavor eaten raw. As a full-sized tree, basswood can produce a prodigious amount of edible greens, albeit for a short window of time. Perhaps the most noted use of basswood is its extensive utility in the production of cordage (i.e., bast, its namesake) for uses as diverse as basket making, rope, mats, and fishnets.

Basswood is often considered an indicator of high pH sites and has a high calcium requirement. That said, I have seen it sporadically in infertile, high-elevation areas with chestnut oak and heaths; like other trees, it can be plastic as to habitat requirements. Its leaves contain high levels of calcium, magnesium, nitrogen, potassium, and phosphorus, and decay readily into high-pH humus. On 30 sites in the Adirondacks, Michael Kudish found the median pH under *Tilia* to be 5.8.[359] Basswood is moderately shade tolerant and is often found in the subcanopy below other trees.

Basswood (*Tilia americana*).

Tripsacum dactyloides • Gamma Grass

Great-grained grass holds permaculture potential and popcorn potency.

This large grass can grow nearly ten feet tall, bearing spikes of large, widely spaced seeds. These can be popped for popcorn or used for grain; they contain up to 27% protein. Researchers at the Land Institute in Kansas have investigated the possibility of developing a perennial native grain crop from the species.[360] Gamma grass is related to corn but immune to several of corn's agricultural pests.

Gamma grass is found in sunny floodplains, marshes, and in coastal habitats. Farther west, it is a tallgrass prairie species found in moist, rich, open habitats and edges. It is the host plant for several moth and butterfly species.

Tsuga canadensis • **Eastern Hemlock**

Evergreen tea tree.

A hemlock grove casts a distinctive purple shade, as the deep green foliage filters light hitting the castoff needle duff below. Hemlock is a long-lived species, living 400–600 years, and in suitably moist, cool habitats, shady groves composed primarily of hemlock can predominate. It is most frequently associated with yellow birch and sugar maple.[361] Very shade tolerant but not drought tolerant, young hemlocks grow best in the shade of other species. Groves of mature hemlocks cast deep shade year-round, and maintain a highly acidic duff beneath them. Therefore, the understory flora under hemlocks is usually scant.

Hemlock seeds are consumed by chickadees, pine siskins, goldfinches, juncos, crossbills, and squirrels, and hemlock groves provide critical nesting habitat for spectacular Blackburnian warblers, as well as black-throated green warblers, and golden-crowned kinglets.[362] I remember hiking up to Giant Ledge in the Catskills just as the male Blackburnian warblers arrived in spring. They were in full song and their golden orange throats glowed against the dark green foliage of the hemlocks.

Hemlock woolly adelgid is an exotic insect pest from eastern Asia that arrived in eastern North America in the 1950s. It is killing hemlock trees over the majority of their range. Recent bio-control innovations may offer some hope for this majestic, long-lived tree species.

We primarily use hemlock as a tea. It has the distinctive tang of the coniferous trees, grounded by mellow, rich, nourishing bass tones. It is high in vitamin C. The inner bark of the tree is astringent and has been used for colds by a variety of northern Indigenous peoples including the Iroquois, Micmac, and Algonquin.[363]

Typha latifolia • **Broadleaf Cattail**

Marsh grocery.

A prominent species in the grocery store of the marsh is broadleaf cattail. The dark punks and tall leaf blades of this plant are a familiar sight to anyone who has even driven by a wetland let alone spent time wandering in the marshes and shallow pond edges where this species reaches its greatest extent. Even a wet ditch can house its share of cattails, and that turns it from a normal ditch into a swale full of food (provided, at least, that the ditch water is clean).

Cattail abounds in edible parts and cultural uses. My favorites are its "hearts," the white peeled cores at the base of stems before flowering, and its golden pollen, which is highly nutritious and can augment other flours in baked goods. Others speak highly of the young staminate spikes eaten like corn on the cob, but I always miss the timing and have yet to try it. Some mention that starch can be extracted from the roots, but many say so ruefully, joking about the effort involved. However, there it is, in great abundance — flour of the marshes.

Other uses for cattail abound. Cattail leaves are a staple of Indigenous weavings. The punks aloft on the stems can be dipped in tallow and made into torches. Lacking handy tallow, I've done the same with spruce resin, which makes for a sputtery, violent torch but good entertainment for pyromaniacs and their bold children. A more gentle child's pastime with cattail punks is to loosen a chunk of the fluffy seeds and then run in the wind, the punk disgorging thousands of parachuting propagules in its wake. It is good to know these things if bringing youngsters to the marsh.

Ulmus rubra • Slippery elm
Scratchy leaves, soothing bark.

Slippery elm is a tree with big, scratchy leaves that feel like cats' tongues. In contrast to its scabrous leaves, it has a soothing inner bark that makes a highly nutritious mucilage. It can be used to sooth areas of irritation, and is also a very gentle food for convalescents who need to slowly rebuild their vitality. Its soothing properties can be used externally, for shingles, erysipelas, and other skin inflammations, as well as internally for coughs, sore throat, diarrhea, and irritable bowel syndrome.[364] Tommie Bass reported, "The way to use the

bark is to strip it from the tree and dry it. Then you can pound it until it is a powder and make a soup out of it."[365]

Slippery elm is less susceptible to Dutch elm disease than its congener, American elm. If it is sited in a place where eventual mortality might be an issue (such as near a house or fence), consider coppicing it as a multistemmed tree and harvesting stems periodically for medicine or other uses.

Slippery elm thrives in fertile bottomland soils. When present on drier uplands, it is a good indicator for calcium- or magnesium-rich soils. Its seeds are a food source for many bird species (from wood ducks to rose-breasted grosbeaks), and many butterfly and moth species host on slippery elm as well.[366]

Vaccinium Section Oxycoccus: The Cranberries
Late autumn sour fruits.

Cranberry (*Vaccinium macrocarpon*) is a species of mountain and coastal bogs and marshes. Its seeds often germinate within the evenly moist matrix of sphagnum mosses. It spreads by horizontal stolons and can form a dense, low mat.[367] Its fruits are the familiar, sour fruits consumed by Americans primarily around Thanksgiving time. Like other *Vaccinium* species, cranberry is adapted for pollination by bumblebees and other native bee species.

Cranberries "contain proanthocyanidins that have been shown to block the adhesion of *Escherichia coli* to human urothelium in vitro and in clinical trials" and are useful in the treatment of urinary tract infections. Several other *Vaccinium* species have shown promise in this regard as well.[368]

Small cranberry (*Vaccinium oxycoccus*) is seldom cultivated or harvested, but can be found inhabiting northern bogs.

Lingonberry (*Vaccinium vitis-idaea* **ssp.** *minus*), is the New World subspecies of a circumboreal species. It ranges across the Canadian Arctic and south to New England. Lingonberries are eaten raw or cooked and are processed into a variety of preserves, syrups, candies, juices, and wines. Lingonberries are popular in Scandinavian countries and are also harvested commercially in Canada and Alaska.[369]

Lingonberry habitats include openings such as shores, ledges, and meadows, as well as northern coniferous forests. They occur in both wetland and non-wetland habitats.

Vaccinium Section Cyanococcus: The Blueberries

Wild blueberries of all shapes and sizes.

Sunny hills in Maine, Newfoundland and Labrador are swathed in **lowbush blueberry** (*Vaccinium angustifolium*), where people dating back many centuries have maintained vast "blueberry barrens" using fire. Primarily a northerly species, it is occasional in the higher elevations (probably 800 feet and above) here in New Jersey, where it is largely replaced by the similar, broader-leaved *Vaccinium pallidum*, also commonly called lowbush blueberry. These low blueberry bushes are similar in form, with *Vaccinium angustifolium* more common in northern or high-elevation habitats and infrequent to absent in much of the Piedmont and Coastal Plain.

Both taste fantastic when ripened by the full sun beating down on a rock outcrop or other open site; neither fruit much

Lingonberry (*Vaccinium vitis-idaea* ssp. *minus*).

when shaded. Blueberries rebound quickly from underground rhizomes after wildfires and intentional burns and are stimulated to grow vigorously in the aftermath of fire.

Velvetleaf blueberry (*Vaccinium myrtilloides*) is another northern blueberry species, mainly found in moist soils and high-elevation open upland sites, where it can intermingle with lowbush blueberry. The fruits, while quite tasty, seem the tiniest bit less choice to me than *V. angustifolium* or *V. pallidum*.

All the blueberry species are bee pollinated and their bell-shaped flowers constitute an important early-season forage for these native pollinators. They bear wonderful, small, sweet fruits that pack in abundant flavor as well as a healthy dose of antioxidants. Tommie Bass recommended a tea of *Vaccinium* species leaves for high blood pressure and diabetes.[370]

Lowbush blueberry (*Vaccinium angustifolium*).

Highbush Blueberries

Those unfamiliar with the wild **highbush blueberry** (*Vaccinium corymbosum*) will be impressed to know that this recent ancestor of the domestic blueberry is a dominant component of shrub understories in wet, acidic soils, and is a charismatic species with eccentric branching, evocative peeling bark, and stunning crimson fall color. Add to this deer resistance and flowers that support early season bumblebee activity, and you have a fantastic native shrub. Not to mention the delicious fruit.

The fruit of highbush blueberry will be familiar to almost everyone, as it is the recent wild antecedent of much of the domesticated blueberry crop (at least those deriving from North America; winter blueberries from South America, and other imports, are from a related species). Domesticated highbush blueberry shrubs were first selected from wild stock in the New Jersey Pine Barrens, and some agricultural cultivars still originate in the sandy, acidic swamps of the pinelands.

In addition to being delicious, the fruits are nutrient dense and include vitamin C, vitamin K, vitamin B6, folate, potassium, copper, and manganese. Blueberries are a rich source of the antioxidant anthocyanin, with many associated health benefits.[371] Matthew Wood describes blueberry as "excellent for thin hypoglycemics as a source of immediately ready blood sugar that will not cause a sudden sugar spike and drop, while it is also beneficial for diabetics and people with hyperinsulinism, who need to keep their blood sugar levels regular."[372]

In addition to its fruit, blueberry leaves can be harvested to make a delicious, mild tea that also confers many antioxidant benefits and has been investigated for ameliorating insulin resistance.[373]

Blueberry's bell-shaped pendant flowers are evolved for pollination by bees. One study documented visits by 42 different bee species visiting highbush blooms.[374]

A very similar species known as *Vaccinium caesariense* is known primarily from the Atlantic Coastal Plain. It is distinguished primarily by smaller flowers and leaves.

Verbena hastata • Blue Vervain
A blue flower for the blues.

Some flowers look best in masses. Blue vervain is one of those: its spikes bear only thin rings of blue flowers, but the effect of dozens or hundreds of these spikes borne aloft over a wet meadow is entrancing.

Blue vervain is an herbal nervine: a calming plant that is useful for treating a wide range of stress-related ailments. It is especially suited to people who are driven, self-critical, suffer from nervous exhaustion, and "can't shut off," in the words of herbalist Lise Wolff.[375] It also serves those with poor circulation and women with difficulty during their menstrual cycles. Matthew Wood writes "there is probably no better remedy than blue vervain for menopausal night sweats and hot flashes." [376] Darryl Patton adds it to cough formulas as a bronchial nerve sedative.[377]

The Viburnums
A shrub for every habitat.

The viburnums are a widely-distributed genus of shrubs and/or small trees that inhabit nearly every habitat in our range. They generally produce edible fruit (though not all viburnums do so). Blackhaw and highbush cranberry are used for the antispasmodic properties of their inner bark.[378]

Hobblebush (*Viburnum lantanoides*) is a low, broad shrub with beautiful wide leaves (especially in late summer and fall when they take on a rich palette of colors.) This is a shade-tolerant understory species at home in highly acidic soils, and is usually found in the northern parts of our range. Hobblebush has edible fruit that are described as sweet and "best after a frost." [379]

Nannyberry (*Viburnum lentago*) bears soft sweet fruits borne on tall woody stems. Nannyberry is a multistemmed understory tree that bears an edible drupe with no direct analog among domesticated fruits. The fruits resemble a small, purple olive. They contain a single seed surrounded by a sweet, soft pulp. The flavor is somewhat like dates or prunes, though less sweet. Both the fruit and the tree are reminiscent of blackhaw (see below). However, nannyberry produces larger fruit. Nannyberry is somewhat eccentric in form, with long wayward stems radiating from bent trunks.

Withe rod (*Viburnum nudum*) is a widespread shrub found in swamps, bogs, and occasionally in open, moist forests. It too has edible fruit. Like its relatives hobblebush, blackhaw, and nannyberry, its fruits are reminiscent of raisins, dates, or prunes, and an alternate name for the species is wild raisin.

Blackhaw (*Viburnum prunifolium*) is a small tree (or large shrub) and a common component of shaded woods, from mesic forests to wetland edges, and in open rocky glades over rich geology. It looks a bit like flowering dogwood due to its corky, platey bark, but tends to have a more unkempt, eccentric appearance. It bears abundant clusters of small flowers in spring, has wonderfully fiery fall color, and produces a late-season fruit in October when many other edible fruits are already exhausted. The fruit is soft, somewhat prune-like in texture, mildly sweet, and has a good, tannic finish like Oolong tea. We either eat it fresh or puree it and use it for cakes. I enjoy the flavor but usually eat a limited amount

Blackhaw
(*Viburnum prunifolium*).

of fresh fruit, because each small fruit has a seed that needs to be spat out and is simply more time consuming than snarfing down most larger or pitless fruits.

Blackhaw makes an antispasmodic medicine with nervine and tonic properties.[380] Appalachian herbalist Tommie Bass utilized blackhaw for menstrual cramps and sometimes as an ingredient in his arthritis tonics.[381]

Highbush cranberry (*Viburnum trilobum*) is a tall shrub found along river banks and other moist edges, conspicuous due to its bright red fruit. Unlike its namesake, it is not a heath but a member of the genus *Viburnum*. Like cranberry, *Viburnum trilobum* can be used to produce jellies, sauces, and wines — the raw fruit is exceedingly sour.

Highbush cranberry is closely related to a European *Viburnum* species often called highbush cranberry as well, or crampbark by herbalists. This is *Viburnum opulus*, and the two are so closely related that the native *Viburnum trilobum* is sometimes known as *Viburnum opulus* var. *americanum*, or *V. opulus* ssp. *trilobum*.

Why the tangent into taxonomy? Because the European species can also be found in our region, naturalized in the wild landscape. It is uncommon in North Carolina, Pennsylvania, and New Jersey, but found in almost every county in northern New England.[382] Meanwhile, the native species is rare in Pennsylvania and New Jersey.

The European species, while useful medicinally, has a distasteful fruit. According to forager Sam Thayer, the native species is "highly esteemed for [its] fruit, while the introduced *V. opulus* has terrible fruit."[383]

The morphological difference between the two is fairly subtle, but is crystal clear once you look. The leaf stem (petiole) of both species has a few raised glands. The European species has wide glands that are like pitted lumps on the leaf stem; the glands are convex or at least truncate at the top. The American species has stalked glands that are concave, and the glands are taller than wide.[384] I have seen the European species for sale even at native plant sales, so it is best to carry a magnifying glass or loupe and verify the identity of any wild or commercial specimens for yourself.

Highbush cranberry is used herbally in the same way as its European relative, for the relief of cramping, including menstrual cramps, stomach cramps, and musculoskeletal discomfort.[385]

Viola sororia • Common Blue Violet
Pretty little weed makes medicine and pretty good food.

The pretty little violet found in lawns along with dandelions and other weeds is a native species. It ranges from fields to forest habitats, and can be somewhat variable in expression, leading some botanists to split it into several species. It has edible foliage and its flowers can be used as an edible garnish or to color a beverage. Violet leaves contain high levels of vitamins A and C, as well as soluble fiber. The constituent rutin, a glycoside of quercetin, is antioxidant and anti-inflammatory.[386]

Violet is cooling and moistening, and infusions are used to soothe coughs as well as dry, irritated skin conditions. It is a lymphatic tonic and is traditionally used in anticancer formulas.[387]

Other blue-flowered violet species are edible too, but this species is so common and adaptable that it seems the most logical to utilize. In forested wetlands, *Viola cucullaria* could be substituted.

Fritillary butterfly larva host on violets, feeding on foliage at night to avoid detection by birds.[388]

Vitis: **The Wild Grapes**

Fruit loops.

Though wild grapes can be found in many habitats, they usually recruit in sunny, disturbed conditions and grow along with the woody plants — trees or shrubs — that thicket up at the same time. They are woody plants that don't support their own weight, and are thus known technically as lianas, as opposed to vines, which are herbaceous.

A number of wild grape species are common in our area, including *Vitis aestivalis, V. riparia,* and *V. vulpina.* While all these species bear edible fruits, most are best suited for juices and other processing. **Fox grape** (*Vitis labrusca*), on the other hand, produces large, sweet, seedy grapes that can make for choice eating.

In addition to its fruits, grape leaves are edible, as those familiar with stuffed grape leaves can attest. When cut, grape "vines" drip sap that can be used as an important source of fresh water in a survival situation.

Sometimes foresters and other land managers decry grape vines for harming trees. However, these are not choking lianas like wisteria or Oriental bittersweet, and they have high wildlife values. Wild grapes are consumed by pileated woodpeckers and other frugivorous birds as well as a variety of mammals.

An early account of Indigenous cultivation of a native species involves grapes: "In one of the earliest reports of conditions in the New World, Verrazano noted, of wild grapevines on what is now the Delmarva Peninsula, that 'the bushes around them are removed so that the fruit can ripen better.'"[389] To this day, fox grapes are cultivated as Concord grapes, which is a cultivar selected from the species.

Scuppernongs and **muscadine** (*Vitis rotundifolia*) are wild grapes with a southerly distribution and a flavor that is very choice for both fresh eating and processing.

An intriguing account by the Spanish priest Juan de Carrera from 1572, and quoted in William Doolittle's *Cultivated Landscapes of Native North America,* describes the cultivation of scuppernongs by native people in the Chesapeake Bay area: "... a very beautiful orchard, as well laid out and ordered as the vineyards of Spain. It was located on sandy soil, and the vines were laden with fair white grapes, large and ripe. Also within the vineyard there was a great number of plum, cherry, and persimmon trees." [390]

Zizania aquatica • **Wild Rice**

Our native rice.

Wild rice is a native grain borne on a grass that thrives in still waterways, often forming extensive, pure stands. Indigenous peoples including the Chippewa, Ojibwa, and Potawatomi [391] use the grain for flour, to thicken soup, and with game; [392] its use is most prevalent in the Great Lakes bioregion but the species ranges throughout the Northeast as well, primarily in the Coastal Plain. Nowadays, it is harvested and sold as a gourmet food. It is highly nutritious, with a high protein and caloric content.

Wild rice is a reseeding annual, renewing itself every year from grains that shatter from its tall seedheads and settle into the mud. It can grow to a height of ten feet in its single year of growth.

The species is tolerant of somewhat brackish water. Its seeds are readily consumed by waterfowl.

Wild rice can be intentionally introduced in ponds and other inundated habitats with slow or still water. It will grow in water to about four feet deep, though it prefers depths of around 18 inches to two feet.

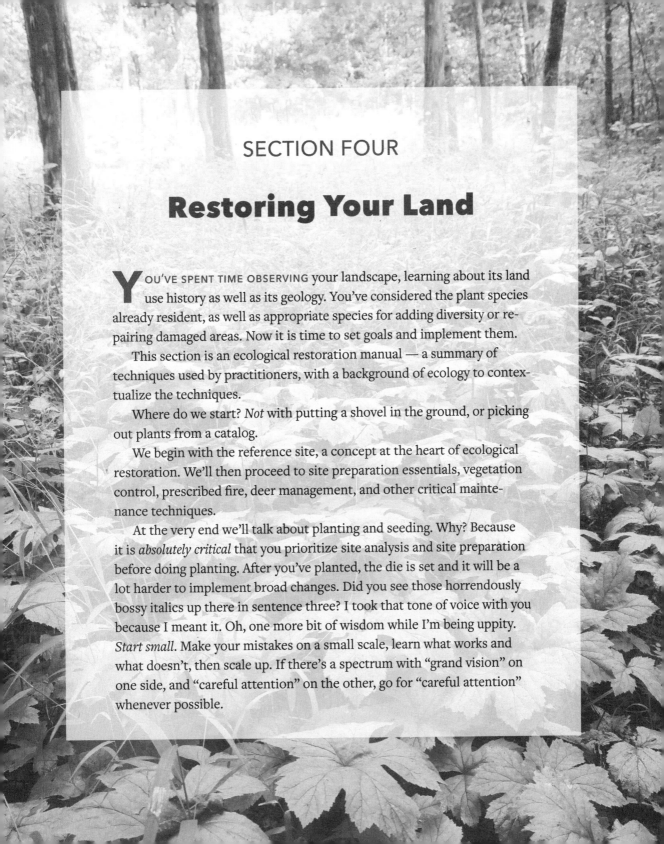

SECTION FOUR

Restoring Your Land

YOU'VE SPENT TIME OBSERVING your landscape, learning about its land use history as well as its geology. You've considered the plant species already resident, as well as appropriate species for adding diversity or repairing damaged areas. Now it is time to set goals and implement them.

This section is an ecological restoration manual — a summary of techniques used by practitioners, with a background of ecology to contextualize the techniques.

Where do we start? *Not* with putting a shovel in the ground, or picking out plants from a catalog.

We begin with the reference site, a concept at the heart of ecological restoration. We'll then proceed to site preparation essentials, vegetation control, prescribed fire, deer management, and other critical maintenance techniques.

At the very end we'll talk about planting and seeding. Why? Because it is *absolutely critical* that you prioritize site analysis and site preparation before doing planting. After you've planted, the die is set and it will be a lot harder to implement broad changes. Did you see those horrendously bossy italics up there in sentence three? I took that tone of voice with you because I meant it. Oh, one more bit of wisdom while I'm being uppity. *Start small.* Make your mistakes on a small scale, learn what works and what doesn't, then scale up. If there's a spectrum with "grand vision" on one side, and "careful attention" on the other, go for "careful attention" whenever possible.

CHAPTER 10:

The Reference Site

AT THE HEART OF ecological restoration is a humble precept: look to the existing indigenous ecosystem for a model of what a repaired ecosystem should look and feel like. A humble precept, because it is often our tendency to privilege the newest, grandest master plan, the most radical intervention. In ecological restoration, we recognize that the human ecological imagination is limited, that much of ecological complexity is outside our grasp or design, and that the best way to honor life's complexity is to replicate systems where we find it already.

Ecological restoration is anchored to the reference site — whether it is the prairie, the oak woodland, the dunescape, or the serpentine barrens. By contrast, novel ecosystems dreamt up by a designer might look like habitats — with aspects of ecological function, gardens of plants, some roving wildlife or domesticates — but they are an artificial flavor, a synthetic derivative of the most obvious characteristics of a natural system but lacking its complex soul, the suffusion of relationships that is beyond our direct artifice.

Our goal is not to design wholly new systems, but to allow the existing patterns of life of the planet to cohere again in places they have been ousted from. Get good enough at that and then we can talk about novel ecosystems, colonizing asteroids, tampering with the genome, and all that grandiose stuff.

How to Find a Reference Site

Reference sites give us an idea of the potential of our restoration site. We visit these special places to understand what kind of plants and animals can live in our site, as well as the relative frequency, density, height, cover, and phenology of species.

You will likely be choosing from among local natural areas to find your reference sites. State and county parks, nature preserves, and urban parks will be among your first choices. While it is always nice to see a "pristine" example of local nature, make sure to also visit sites with similar land use and disturbance histories to your own. Each will have something to teach. Here are a few other ways to find good reference sites:

- Inquire with your local native plant society or conservation group. Explain what you are doing and you may find that experienced naturalists are eager to help.
- Botanical societies often have field trips led by members. These are usually to exemplary sites, and often span the urban/rural spectrum. Journals of these societies, such as *Bartonia*, the *Journal of the Torrey Botanical Society*, *Rhodora*, and *Castanea* all have at least some regional focus and contain articles such as inventories, field trip notes, and monographs. Old journal issues are especially fascinating (and heartbreaking) because they reveal how diverse sites once were that are now much less so.
- Plant lists from botanical surveys are archived online at UniversalFQA.org (more about this in the Appendix). Because these lists have floristic quality metrics associated with them, they represent an ideal resource for identifying high-quality sites. Note that lists are accessed by

Exploring a north-facing slope over amphibolite rocks.

site name, and may be more or less opaque as to location and applicability.

- The Office of Natural Heritage for each state tracks rare species occurrences. They also often have resources and publications detailing local plant communities, and they may (or may not) have helpful staff willing to clue you in to worthy reference sites in your area.

Interpreting a Natural Area for Beginners

There can be so much complexity in even a small patch of the natural world. Who's making that call or song? What flower is that? How do I identify rocks? How the heck does water flow when it's under the ground?

Faced with the unknown, we have two choices — get frustrated or get psyched. I'm advocating for the latter. There's so much out there to learn, and the best possible way to learn it is in direct interaction with the land. The open field of the unknown is fertile ground for those of us who hunger for new sensations, new challenges, and ultimately, a much richer experience of life.

Hepatica.

Meanwhile, you need to interpret reference sites, as well as your own project site. You may want to know everything before you start, but it's not going to work that way. Start small on your project site. Focus on one type of natural entity. Get good at plants ... and birds, pollinators, and soils can follow.

So you've found a dynamite reference site. It's clearly beautiful, but you can't tell a maple from an oak. I hear you. I was there, and not that many years ago. I remember finding white oaks in the woods my first winter living in a rural area. I had a little field guide Tree Finder, and was trying to ID trees using their bark, since most of the leaves were on the ground. I guessed that the white oaks were silver maples, because, well, the bark was kinda silvery.

I was set straight later that year by a chance comment from a local guy who was helping us get our little two-wheel-drive four-banger pickup truck unstuck. From next to a white oak. So begins the career of a botanist.

Luckily, plants are a lot easier to identify in the growing season than in the winter, and they space themselves out in a regular way, like classes in a school course. What do I mean? Plants are most readily identifiable when in flower. They flower through the season. On any given day, only a few new species (at most) will be flowering unless you're really lucky. Take regular walks in natural areas, and try identifying what's newly in bloom. I'm going to put in a pitch for doing this the old-fashioned way, with a field guide like *Newcomb's Wildflower Guide*. After all, you're seeking immersion and sensual experience, not more screen time or a list of names. However, there are myriad apps that will do your ID work for you. If you use one, try to spend some time with each plant getting to know its feel, scent, and textures. Hikes guided by a local naturalist are also a good way to start, and if you can't access them all the time, you can still prowl their Instagram pages to see what they are seeing.

A lot of garden and landscape books are like self-help books and fad diets, guaranteeing you stunning results in as short a time as possible. I'm not gonna con you. Interpreting the natural world is a lifelong process. And ecological restoration is still young. We are all beginners, in this together.

Now let's put a shovel in the ground.

Steep forested cove reference community.

CHAPTER 11

Repairing Soils

THE PRODUCTIVITY OF SOILS can be improved, and damaged soils can be repaired. Many soils are degraded by agricultural or industrial land use, and may lack appropriate ecology in the form of bacteria, fungi, and other soil organisms. Degraded soils also lack ideal structure, either due to mechanical effects such as compaction and plowing, or because of the extirpation of the soil fauna and fungal community which aerates and structures soil, forms larger soil aggregates out of smaller materials, and cycles nutrients.

Soils can often be improved through the considered addition of appropriate organic matter. In some cases, soil improvement may involve the introduction of soil organisms. If the soil is thought of as a living being analogous to our body, adding different types of organic matter is analogous to altering or augmenting one's diet. To spur soil life, we can add materials that are "prebiotic" — the chosen foods of various soil organisms. For example,

many beneficial soil fungi will thrive on high-lignin amendments such as wood chips. It is also possible to use "probiotics" for soil — directly introducing desirable soil organisms by means of compost teas or fungal spawn.

Soil Amendments

Wood Chips

One combination we've worked with to repair damaged soils on small areas is to top dress with hardwood chips, and inoculate the chips with sawdust spawn of wine cap mushrooms (*Stropharia rugosoannulata*). The bed will fruit out with delicious edible mushrooms while the mycelia of the mushrooms meld the chips and underlying soil into a wonderfully fertile, well-structured growing environment suitable for the cultivation of native woodland species.[393]

A mulch of wood chips should be one inch thick or thinner if you intend to plant into it immediately. For larger area applications, some percentage of the soil can

be left bare, other areas piled more deeply
with wood chips, to encourage heterogeneity and thus diversity. For example,
large seeds like tree nuts germinate well in
thick piles of wood chips (and rodents like
to dig in wood chips to cache them) whereas other, smaller seeds will only recruit on
exposed soils.

Leaf Compost

Native plants generally do not require
intensive fertilization. However, the creation of a rich, fertile soil can be beneficial,
especially for many woodland plants, which
evolved in the context of a biologically rich
and moisture-retentive soil.

Leaf compost is a practical amendment
for preparing small-scale woodland plantings or augmenting the productivity and
growth of edible plantings. Incorporation
of fully composted leaves into the existing
soil to six inches depth is ideal only if you
are preparing beds in an area where desirable plants are absent. If soils (and roots)
are to be left undisturbed, areas can be
top-dressed with leaf compost instead. A
thin layer of raw leaves can be used instead,
or a thicker layer if some weed suppression
is desired.

Mycorrhizal Fungi

Fungi that form direct associations with
root and other tissues of plants are known
as mycorrhizal fungi. These fungal species
aid plants in obtaining soil nutrients and

Wine cap mushrooms (*Stropharia rugosoannulata*) in
inoculated planting bed.

water, and can be many times more effective and expansive than the plant's roots
themselves. Many specialist plant species
are heavily dependent on these mycorrhizae, and have root systems with few fine
feeder roots. On their own, these plants
will fail to thrive.

Agriculture destroys or greatly diminishes the presence of these fungal partners.
Tillage chops mycorrhizae to myriad pieces

and exposes sensitive fungal hyphae to the sun. The lack of perennial plant partners in an annual crop field means that remnant or recruiting fungal fragments or spores are often unable to thrive or grow.

Arbuscular mycorrhizal fungi don't create aboveground fruiting bodies (mushrooms). Because they are limited to belowground spore production, it can be very difficult for these species to colonize new, non-contiguous sites.

If you're restoring a post-agricultural field or forest, your site may benefit from the addition of mycorrhizal fungi — a probiotic, as it were. One way to accomplish this is to find a high-quality reference site and obtain permission to transport small amounts of native soil from there to your restoration, incorporating these by hand. More effective may be to inoculate native seedlings with mycorrhizal spores, either purchased from a specialist supplier of native fungi or by growing the seedlings with a small addition of soils from a remnant. Once these seedlings mature and are fully colonized by fungi, they can be planted out on the restoration site as "nurseries" for beneficial fungi.[394]

For the time being, commercial sources of native fungi spores are rare. However, those working at large scales will want to consider this option and seek out sources. Remnant natural habitats have enough threats to them without adding widespread soil harvesting for restorations elsewhere!

Biochar

Biochar is a long-lasting form of organic carbon that holds water and soil nutrients in its matrix. Because it is recalcitrant, biochar can sequester carbon into soils for hundreds to thousands of years. Its structural characteristics allow it to provide ideal habitat for beneficial soil fungi and bacteria. Biochar is the same thing as charcoal — lignified biomass smoldered in the relative absence of oxygen through a process known as pyrolysis.

The inspiration for biochar comes from the rediscovery of centuries-to-millenia old highly fertile soils in the Amazon, a region known for extremely poor soils. These are found near historic settlement sites, and it has been theorized that these soils, called "terra preta," were created by pre-Columbian Indigenous civilizations. The process for creating terra preta is poorly understood but probably involved a combination of burnt agricultural refuse and some as-yet poorly understood biological process or entity which imbued the char with exceptional soil life and long-lasting fertility.

Decompacting Soil

Soils can become compacted as a result of human or vehicular traffic, construction equipment, or the long-term use of agricultural equipment, especially plows that work the soil to the same depth every year and create a hard, impervious plow layer below. Some methods to reduce compaction include:

Hand excavation

This involves carefully digging out soil, loosening it, and replacing it, sometimes with amendments such as biochar or organic matter to encourage structure and soil organisms. Freshly decompacted soils on slopes should be protected by a layer of mulch or a soil blanket of jute or coir to prevent erosion.

Vertical staking

In this method one drives many wooden branches or stakes into the soil, physically separating it but also creating porous spaces when the introduced wood rots.

Subsoiling

In the case of subsurface compacted layers (like a plow layer), specialized tillage equipment can cut through the impervious layer at intervals and allow drainage to occur.

Note that in areas already vegetated with desirable plants, the process of decompacting soils through digging or amending can be very harmful to existing root systems. In this case, vertical staking may be a better option.

Restoring Landscape Structure
Pit and Mound Topography

Most agricultural lands have been flattened to improve efficiency in plowing, planting, and management. Rocks have been removed and irregularities corrected. The resulting fields are far more homogenous

Land shaping.

Tip-up mound with Christmas fern (*Polystichum acrostichoides*) and unidentified moss.

drain wet but fertile lowlands for agricultural use.

Remnant wild landscapes often manifest a much more heterogeneous topography and hydrology. In mature forests, the exposed root systems and trunks of wind-thrown trees create their own small hillocks, with a miniature valley below where the tree has heaved out. This is called pit and mound topography, and it results in a gradient of soil moisture and shade from the sheltered, moist pit to the dry, exposed mound. This pattern is repeated over and over through the forested landscape, providing for a diversity of plants adapted to differing conditions. Moss and fern spores often germinate on the exposed soils of tip-up mounds, especially those shielded from direct sun. Topographical diversity also provides a barrier to the monopolization of an entire site by just one aggressive species, as all species may encounter specialized conditions of wet or dry to which they are less adapted.

In addition to promoting plant diversity, the restoration of pit and mound topography can protect soils, and help infiltrate water. In the hilly and mountainous regions of the Northeast, colonial-era agricultural conversion led rapidly to soil erosion as rainstorms gushed and then gullied down homogenous, sometimes bare slopes. To encourage water to infiltrate rather than flow over the surface, a series of pits and mounds (sometimes

than they once were. Streams have been rerouted out of the center of valleys to their peripheries to increase arable land and dispense with water-imbued wetlands, and ditches and tiles have been installed to constructed as swales and berms) can slow the movement of water and the transport of soils and encourage infiltration.

Structural Repair Methods

Slope, Bank, and Gully Stabilization

Several specialized techniques aim to stabilize erosive areas subject to moderate to intense water flows. Generally speaking, our aim with water is to slow it, spread it, infiltrate it, and restore the full hydrological potential of a site.

These methods hold in common that they often combine an engineered soil structure (dam, berm, etc.) with living plant materials that will hold it and ultimately anchor the new structure. These structures must be used closely together, relieving small amounts of gradient, or these low-impact stabilization techniques will fail or blow out. Construction of these structures should progress from the bottom of a slope up, or downstream upwards in the case of check dams.[395]

Check Logs

These are logs placed on contour and staked in place on moderately erosive slopes. They can slow water flow and capture sediments.

Live Stakes and Fascines

These are cuttings of easy-to-root native woody plants that can be thrust into exposed and erosive steam banks. They will root deeply, especially if "planted" into the soil at a 45 degree angle. Many wetlands species are able to function in this way, including willows, elderberry, buttonbush, silky dogwood, and some viburnums. Fascines are bundles of cuttings that provide structural mass, slowing water or soil while rooting in. Live stakes can be used to root and hold in place engineered structures such as coir logs.

Check Dams

These are low dams built into a scoured, channelized stream corridor and placed at close intervals along a sloping reach. They slow water speed — they aren't designed to fully stop floodwater, as with an impounding dam, but to check its flow velocity by causing water to have to rise to a level over the dam and then overflow to the gradient of the next dam. In the process of slowing water flows, sediments settle behind each check dam and raise the floor of the previously scoured channel closer to the grade of the bank. By raising streambeds in this way, floodwaters can once again reach stream floodplains rather than rushing off-site to compound flash flooding issues.

CHAPTER 12

Vegetation Control

AKEY ASPECT OF SITE PREPARATION involves the removal of undesirable plants from the vicinity of the restoration area. Undesirable plants may compete with existing, desirable species, or with restoration plantings. For example, if you're converting a lawn to meadow, existing Eurasian cool-season lawn grasses may need to be removed or managed for. These turf grasses are highly competitive with meadow seedlings for moisture, fertility, and root space.

Several site preparation methods for removing large swaths of vegetation are summarized below.

Tillage

The use of a cultivator or tiller can rip up existing vegetation and root systems. Tillage is primarily relevant for undesirable herbaceous vegetation in a nonwoody-plant-dominated system, such as a lawn or open field.

Tillage provides a loose soil for planting. It helps with incorporating soil

amendments. A disadvantage of tillage is that it brings dormant weed seeds to the surface and triggers their germination. It can also ruin soil structure, especially if soils are too wet or too dry or tillage is chronic. The mycelium of sensitive fungal species are intolerant of sun exposure and can be severely damaged by tillage.

A flame weeder or hoe can be used after tillage, in order to remove the small weed seedlings that germinate in response to soil disturbance.

Smothering and Solarizing

Smothering involves the use of a sheet of heavy black plastic to smother weeds by depriving them of light and elevating temperature levels in the full summer sun. This practice can be done for two weeks after tillage or herbicide use, to suppress new weed germination, or it can be done for a longer time to kill established perennials. For smothering on large scales, some people use silage tarps. We tend to use rolls of

Smothering, at the
Adirondack Wild Center.
CREDIT: RACHEL MACKOW

polypropylene landscape fabric and affix it with ground staples. These are permeable to water and nutrients while still suppressing weeds.

A disadvantage to smothering is the long time period necessary to kill established perennials. In some tenacious species, spans of a year or more may be necessary to kill off root systems.

Clear plastic sheeting can also be used, and can quickly "cook" underlying vegetation, most effectively during the hottest time of the year with the strongest sun. This process is known as "solarizing." We often have old greenhouse film on hand to use for this purpose, and you may be able to score some for free from a nearby nursery.

The "Lasagna" Mulching Method

In "lasagna" gardening, a thick layer of newspaper or other biodegradable paper is placed over existing herbaceous vegetation, usually weeds of modest stature. This paper layer is lightly watered with a hose to keep it in place. A several inch layer of weed-free leaf compost or other compost is spread over the paper. Given enough available biomass, layers of thatch, straw, twigs, or other plant material can be layered under the weed-free layer, creating a "lasagna" of fertile, decomposing ingredients.

This method is well-suited to smothering lawns. It is not recommended for aggressive, persistent perennials that are well-established. It is probably most practical for relatively small sites. This approach can be combined with tillage.

Herbicides

Herbicide use is often the fastest method to remove existing vegetation. With certain established perennials, especially some invasive species, it can be one of the only effective choices. With woody plants, it can be combined with cutting to remove undesirable shrub and tree species.

Herbicides have many disadvantages, including unintended effects on adjacent vegetation, on wildlife, and on the humans applying it. We do not fully understand the long-term effects of various herbicides on the soil biome, but there is mounting evidence that chronic use of herbicides such as glyphosate can be extremely deleterious to soil organisms. However, a limited (not chronic) use of herbicides, applied carefully, may be judicious.

There are both practical and philosophical considerations in using herbicides. We opt into a war paradigm every time we use chemical poisons to alter plant communities. I try to be mindful of that, even in situations where I do recommend their use.

Some species are practically impossible to get rid of without herbicides. Sometimes the soil damage, labor time, gasoline or other resource use involved in *avoiding* herbicides can constitute a significant environmental impact and degrade your site.

Herbicides can be an appropriate tool for invasive species removal in restoration projects, where the targeted species constitute a serious impediment or threat to the establishment of a diverse plant assemblage. If a one-time (or limited time span) use of herbicides will enable the conversion of a site that is otherwise monopolized by an invasive species, I believe the use is justified.

Invasive species best removed with herbicides include lesser celandine, Japanese knotweed, and mugwort — perennial herbs with intractable root systems that will resprout from remnants even if hand-weeded ad infinitum.

Knotweed provides a good case study. In one documented restoration experiment, knotweed (*Reynoutria japonica, R. sachalinensis*, and hybrids) was managed for three years along the Bronx River in New York using two methods: cutting three times each growing season, and cutting plus root and rhizome removal. After three years, knotweed was not eradicated by either method, and the native herbaceous community was not able to recover.[396] Knotweed is a tough species and has even been documented growing through asphalt.[397] That said, control of knotweed with herbicides can be difficult to accomplish as well. There are no easy answers here.

Suckering, large-statured shade-tolerant shrubs such as buckthorn and Oriental photinia are difficult to deal with via mechanical removal. In a forested context, you can't run a dozer through the entire understory and you need other options. Basal bark applications of herbicides, or application of a concentrated dose to cut stumps, may be your best option.

Herbicides can also be useful in creating a "blank slate" before seeding a meadow. This might also be accomplished through repeated tillage, or, on a small scale, smothering — but herbicide remains an economical and time-efficient means. The sheer flowering bounty and diversity of a nicely established meadow seems to me to redeem the one-time use of chemical poisons to remove the lawn or weedy area which preceded it.

On our farm, we don't use any herbicides. Because we live here, we can do the hard work of keeping an eye on everything, and hand-pull, smother, or cut as necessary. Does this make me a hypocrite when I advise the use of herbicides on wild lands that don't benefit from the same level of constant attention, are more highly degraded, or are too large in scale to work by hand? Maybe. There are no easy answers here.

Weeds—When Not to Worry

Many people fear weeds. They carry a cultural stigma, and signify failure in the garden or home landscape. With the growing awareness of invasive plant species, this fear has become almost paralyzing for some.

To dispel that fear, it is best to better understand weeds. Here is a simple two-part perspective on weeds:

1. **Weeds are a symptomatic response to a particular type of disturbance.**

2. **Shift or eliminate the disturbance, and the weeds will not recruit again.**

There's two things here that work in our favor. First, weeds need specific types of disturbance to successfully disperse, germinate, and thrive. Most weeds are species of exposed bare soils. They are annuals or biennials that take quick advantage of resources that are not tied in to the cycles of an established plant community. They grow, seed, and move on, unless the disturbance is repeated.

a newly seeded meadow on bare soil could be overwhelmed by annual and biennial weeds in its first couple of years unless managed with intermittent mowing designed to keep weeds at the same height as native seedlings.

Most of the commonly known weeds are annuals. These are the species that show up in the carefully bared soils of vegetable gardens and in the disturbed soils we create with machinery in the urban and suburban landscape. They include amaranth, foxtail grass, lamb's quarters, purslane, galinsoga, smartweed, and horseweed.

I'd be remiss not to mention that a large percentage of garden weeds are also good edibles. We disturb some soil on our farm every year (with the ostensible purpose of a vegetable garden) in part because of the fantastic crop this yields of amaranth and lamb's quarters, both exceptional cooked greens. If we didn't till the soil in this patch, it would rapidly colonize with asters and goldenrods like the meadow just upslope.

Which leads to the another simple precept about weeds:

Most annuals and biennials do not compete effectively with established perennials.

Right near our "vegetable garden" (AKA annual weed patch) is a small planting of tall coneflower (*Rudbeckia laciniata*) that we maintain for its edible greens. It's specifically desirable because it commences new basal leaf

When not to worry.

Annuals and biennials are primarily a threat to the perennial plants we are nurturing or establishing only when our desired plant community is just a year or two old. For example,

growth very early in the spring. As a green, it is available well before any garden vegetable greens like spinach or lettuce are ready to harvest. It's available well before dandelions, even.

Here's a hypothetical scenario. Let's say one of the lamb's quarters seeds from the garden fell across the path and landed in the tall coneflower patch. While the lamb's quarters are germinating (often late in spring), the tall coneflower has basal leaves almost two feet across, and is starting to bolt for the season. Maybe it's a couple of feet tall already. The lamb's quarters don't stand a chance. Even an annual weed that germinates just as tall coneflower begins its growth in March would be hard-pressed to maintain the growth or access the resources that this well-established native perennial can.

Understanding these two basic precepts about weeds can save us a lot of time and angst as our perennial communities mature. A weed is not a problem sheerly because of its presence, as most of us fear. It is only a problem ... when it is a problem.

The species that are problematic are good to know. Not just by name, but by lifestyle.

In understanding their tactics and duration, it is possible to understand the nature of the problem they present, and a good deal about avoiding, outsmarting, or eliminating them.

For example, Japanese stiltgrass is an extremely competitive weed. It is also an annual. It grows in thick swards sprawling over anything shorter in stature and creating great heaps of thatch that only it seems to be able to germinate through. Rodents like to tunnel beneath its thatch and gnaw on native perennials. Woody plants sometimes suffer from insufficient airflow in swards of stiltgrass.

We mow Japanese stiltgrass right around Labor Day (early September), as it is setting seed. Timed right, it will not have the root reserves to flower again and set seed before the frosts of October. Mowing also chops up the thatch that it would otherwise create, freeing plants and seeds below from its inhibitory effects. If stiltgrass is managed in this way, I've seen new infestations disappear in as little as one year.

CHAPTER 13

Burning

MY COLLEAGUE Kerry and I are running a 100–meter (328-foot) line through some pretty normal looking upland woods in Morris County, New Jersey. Some older chestnut oak, northern red oak, and black oak. Smaller red maple, black birch, and beech coming up. Almost no understory. This park has been devastated by deer overbrowsing and even these more mature forests are stagnant at best. Except something else is going on here ...

We've been here once before, in 2019. Doing the same stuff. Ten one-meter (three-foot) quadrats on a transect, four height classes of understory vegetation, measure the canopy trees.

I face northwest and feed the measuring tape to Kerry. He walks to the nearest canopy tree. Measures the circumference at chest height, shouts it over to me. I tap the digits onto the iPad with cold fingers. It's only March and nothing has leafed out yet. Then swivel to the northeast, and repeat.

In 2019 we collected baseline data in this forest because it's part of an ecological burning program initiated by stewardship staff at the Morris County Park Commission. We're back to see the effects, and they are surprising. Big. The prescribed burn had a dramatic effect on the canopy trees.

In 2021 nearly all the black birch, beech, and smaller diameter red maples are dead. The average distance to the nearest tree using the point-quarter method has jumped from 15 feet to 30 feet. Doubled. The canopy has thinned, and the old oaks are the trees that persisted. Sassafras and blackberry increased, and the pre-existing lowbush blueberry was thriving.

Is this all good or bad? It depends on what the management goals are. Oak forest is declining across much of its range. In the East, the result of over a century of fire suppression has been the replacement of oaks by more shade-tolerant species.[398] At least that's the simple narrative. Ecology is hard to generalize in such sweeping strokes. Nevertheless, what we were seeing in our burn-monitoring transect was something interesting. A ghost, if you will. The ghost of pre-colonial forest conditions, which have often been described as more open, with Indigenous fires creating sunny understories and favoring massive elder oaks in the canopy.

But this is just one location, one sample. Burning in deciduous forests in my home state is still a pretty new thing. We've run a lot of transects for Morris County, but they are almost all baseline monitoring, collecting data in anticipation of future fires. There's much to be learned yet.

Indigenous Fire

The uses of fire by Indigenous peoples in pre-colonial North America were myriad, and the extent vast. Fires managed forest structure, kept prairies and meadows open, cleared travel corridors, protected settlements, reduced biting insects, and increased hunting success. In oak woodlands, fire inhibited acorn weevils and improved the acorn yield and harvest conditions.[399]

Fires were set with attention to seasonality and frequency in accord with management goals. According to Robin Wall Kimmerer and Frank Lake, "Land was burned annually if the intent was to increase game by providing new grass forage. Berry patches were maintained by burning on a cycle of three to five years, depending on the ecology of the target species. A 10- to 12-year fire interval was typically observed around beaver ponds to maximize regeneration of aspen and willows to feed beaver."

By varying fire intensity and frequency, and generally burning in small patches, Indigenous peoples created a mosaic of habitats with differing structure and composition, "ensur[ing] a diversity of plant foods, medicines, game, and materials for the subsistence of the people."[400]

According to Kimmerer and Lake, Indigenous-set burns practically disappeared from eastern forests by the early 1700s, in tandem with the genocide and dispossession of Indigenous peoples. Open forests with mature canopy trees were replaced by dense second growth forests, and meadows, prairies, and savanna habitats disappeared.

Transitions

Before widespread fire suppression became the norm in the 20th century, a period of relatively uncontrolled fires swept the East as logging slash from massive clearcuts left enormous reserves of dry fuel on the ground. Land unsuitable for agriculture was cut for charcoal production and other industries in the 18th and 19th centuries, and ignitions from charcoal-making, railroads, and even stills led to numerous fires.[401]

These uncontrolled 19th century fires informed a United States government policy of fire suppression, and the missions of newly created agencies such as the US Forest Service and US National Park Service in the early 1900s.[402] It was not until 1962 that a committee chaired by Dr. Starker Leopold (son of Aldo Leopold) "took the broader ecological view that parks should be managed as ecosystems," which resulted in the National Park Service recognizing fire as an ecological process in 1968 and paving the way for an era of greater fire tolerance and eventually controlled burning.[403]

The consequences of fire suppression have been immense, including impacts on plant community composition and diversity. Young, densely treed forests accumulate leaf litter, reducing plant diversity especially among species that benefit from additional light, from exposed mineral soils, and from fire disturbance.[404] These more homogeneous forests are also more susceptible to pest outbreaks and disease.

Despite prescribed burning of almost two and a half million acres per year in the United States, "almost nowhere has the use of fire kept pace with or even approached historic levels."[405] The relatively moist, forest-dominated habitats of the northeastern United States and Canada are among the least frequently burned areas in North America.

We're still learning how to do this.

Ecological Potential

In the Northeast, prescribed fire is often used in the service of two similar management goals — the preservation of fire-tolerant oak forests from succession to more mesophytic, shade-tolerant species, and the preservation of pinelands from, well ... oaks. OK, that's a bit crass on my part. What it boils down to is that certain classes of trees (oaks, pines) are relatively more fire adapted than other tree species which might recruit in oak or pine forests in the absence of fire, and proper application of fire has the potential to perpetuate forest types which may otherwise dwindle.

Another important use of prescribed burning is to keep meadows, prairies, and other open habitats from succession to forests.

So fire is often used as a broad-brush-stroke approach to removing unwanted woody vegetation. Part of its appeal is that native plants and even whole communities are adapted to fire, thrive with it, germinate and reproduce because of it. Another part is the scale at which it operates. Imagine hand-pulling five acres of Japanese honeysuckle, or ... just burning it to a crisp. Which brings us to another, very contemporary use of burning: to control invasive species that are less fire-adapted than the native flora.

Understand that the benefits of burning are variable, and sometimes unpredictable. Initial site conditions, existing fuels,[406] seasonality, species composition, and wind conditions are all important variables. A single burn may not change much,[407] or, due to extreme fuel loading, it could have potent consequences.

Burn at Shaw Prairie.

Prescribed fire has the potential to alter not just tree composition, but understory vegetation as well. One study compared unburned and burned oak forest, the latter having experienced 17 years of low-intensity burns. Smaller woody plants decreased 97% and the canopy became significantly more open in burned plots. The abundance and cover of summer herbs increased in burned plots, and spring herbs remained steady. While herbaceous cover increased, native vines and shrubs were lost, and habitat for forest interior birds changed as well. Burning removed most non-native shrubs, but garlic mustard was able to take advantage of exposed soils and seed in prolifically from adjacent, non-burned stands.[408]

Dr. Jay Kelly and colleagues report similar impacts in a 2021 study based here in New Jersey. They recommend that "sensitivity to local contexts and initial conditions is all the more important given the negative effects of prescribed burning observed for various native components of the understory flora, including individual shrub and liana species cover, overall native shrub height and cover, and tree regeneration."[409]

Part of what they are saying is that the effects of burning may be very different on a mesic, north-facing slope versus a dry oaky ridgeline. In the former, typical native shrubs such as spicebush, mapleleaf viburnum, hornbeam, and witch hazel may react adversely to fire. In the oak forest, fire-adapted heaths like lowbush blueberry and black huckleberry will likely flourish after burning.[410] Longer-term applications of prescribed fire might shift moister forests' floras towards more fire- and xeric-adapted ones. At a conference I attended, Dr. Theron Terhune of Tall Timbers suggested that "you need three burn cycles at least to see a shift in vegetation."

Jay Kelly and his colleagues also detail the impact of fire on invasive species in their paper, and there is some really good news here. Cover of invasive shrubs and lianas decreased significantly after two burns, with the most dramatic effects in older forests. Japanese barberry and Japanese honeysuckle were most

negatively impacted, followed by multiflora rose, burning bush, and Asiatic bittersweet.

Putting fire on the ground is an incredibly sophisticated art. I've stood in awe listening to presentation by local fire service folks, and especially the team at Tall Timbers in Florida. A lifetime of knowledge can go into analyzing all the variables pertaining to weather, site conditions, and equipment — not to mention the complex responses of different plant and wildlife species at different seasons and scales.

The impacts of fire are not even limited to what the fire accomplishes by burning vegetation. Soil chemistry will change. Soil temperatures may temporarily shift. A dark, burned soil will heat up faster from sunlight than one covered in leaf litter or thatch. This is used in spring prairie burns to spur native warm season grass growth and disadvantage Eurasian cool season species. But warming is just one potential result. Frank Lake discusses the Indigenous use of the smoke from prescribed burning to block sunlight in hot, droughty conditions and keep rivers cool for salmon! He also says that fire was needed to bring out choice medicinal properties in certain plants.[411]

Fire is an incredible tool and we have much to learn. I have not even touched on safety issues or implementation practices here. For these, a combination of classroom time and field training with experienced mentors is highly recommended.

Mountain laurel (*Kalmia latifolia*) blooming after fire.

CHAPTER 14

Deer Management and Exclusion

WHITE-TAILED DEER are important native animals in intact natural ecosystems. However, the scale of the deer population has exploded beyond the carrying capacity of our remaining wild landscapes throughout much of the eastern United States. Deer overpopulation has had negative impacts on native plants and wildlife, human health, and the health of the deer themselves. Predators of deer such as wolves and mountain lions have been eliminated across much of their former range. Human hunting patterns have changed from year-round subsistence hunting practiced by Indigenous peoples and early colonists, to seasonal hunting with an emphasis on taking trophy bucks instead of does. Forests have been fragmented by development, creating more of the edge-of-woods, suburban, and agricultural habitat preferred by deer. These factors have permitted the deer population to grow to an unsustainable extent. Unsustainable levels of deer browse inhibit tree regeneration

"Crew cut" spicebush stems due to deer overbrowsing.

and decrease the diversity and abundance of native wildflowers and shrubs.

Degradation of native plant communities directly impacts the numerous wildlife species that depend on native plants for shelter and sustenance, dramatically diminishing populations of species ranging from songbirds to butterflies. Deer overbrowse also leads to the invasion of unpalatable non-native plant species, which fill the void left by the disappearance of native plants but do not adequately fill their ecological roles as sustenance and structure for wildlife.

To reduce the deer population, land managers can establish a deer management program on their property. A deer management program includes prioritizing the harvest of antlerless deer, setting target harvest goals every season, and the use of bait stations and deer drives (where groups of hunters "push" deer from hiding places towards a hunting ground). It also must include effective communication with the general public about the ecological need for deer management, and signage and other relevant measures intended to ensure safety.

With deer management, the larger landscape context is key. Neighboring landowners including farmers, large property owners, municipal, county, and state properties, and non-profit conservation organizations should be supported and encouraged in their deer management activities.

Protecting Plantings from White-Tailed Deer

Low, woody growth is a preferred, year-round food for white-tailed deer. In high deer density areas, deer may sample even supposedly deer-resistant plants. Deer

Deer exclusion hoop fence with healthy spicebush (*Lindera benzoin*).

herds in various localities often have different browse patterns and preferences.

Protection of all new plantings is recommended. Deer-repellent sprays or granular repellents, regularly applied, can protect foliage during the growing season. Deer may browse woody stems in winter, causing plants to be disfigured or to perish. Fencing individual plants with welded wire fence hoops is effective year-round.

Manufacturing fence hoops is easy: cut lengths of fence (usually five to eight feet long) and create hoops by folding one end of the fence back to meet the other, forming a circular exclosure or hoop. Use zip ties to hold the hoops together, and ground staples or rebar (attached with zip ties) to hold the fence hoop in place. Small gardens (up to ten feet by ten feet) can also be fenced with four-foot-tall welded wire fencing. Typically, deer do not like to jump into and out of small fenced areas.

In larger-scale exclosures, fences of at least seven feet high are effective. Agricultural fencing of woven metal wire attached to treated wooden posts is a long-term solution. Even this kind of fence needs regular maintenance to remove fallen limbs or undesirable climbing vines. Keeping access open along the perimeter of the fence is crucial, as are regular inspections. Installation of fence in existing forest habitat can be very challenging as installation is often done with heavy equipment such as skid steers, which require broad access

lanes. Some types of lighter fence are being innovated which can be hung from tensioned wires attached to tree trunks. This can be installed with significantly less disturbance. Note that some municipalities may require permits or variances for fencing.

UV-resistant black plastic fencing is available from dealers but I seldom recommend it. It is expensive, and it will sag and be torn by heavy, falling limbs. Black plastic fencing is limited in durability and lifespan, and is not recommended for use under a tree canopy — falling branches will destroy it, though tensioning wire at the top can help repel small falling limbs.

Deer exclusion hoop protecting showy orchis (*Galearis spectabilis*).

CHAPTER 15

Introducing Plant Materials

THE TWO PRIMARY METHODS of introducing new plants to a site are by seeding and by planting live plant materials. Each method has different attributes relative to economy, successful recruitment, and maintenance.

These two methods are not mutually exclusive, but sometimes the site manipulation and maintenance regimens differ enough between the two that one method can interfere with another. For example, taking a nicely seeded bed and digging a hundred small holes in it to introduce live plant plugs might well bury a substantial amount of seed. Likewise, large-scale meadow seedings should be mown at intervals during the early years which might subject established live plantings to unnecessary interference.

Planting

The introduction of live plant materials, in the form of container plants or bare root materials, or even cuttings of reliable species, has the intrinsic advantage of placing a plant where one wants it, and being assured of its existence. This is in contrast to seeding, where recruitment can be spatially haphazard and germination is not assured.

While survivorship can be excellent, plantings of live material have a water requirement, and unexpected drought can pose a significant threat. Ideally, plantings are accompanied by some ability to water them sufficiently to prevent their demise, at least until they are well enough established and rooted to fend for themselves.

There are a few ways to hedge against the threat of drought. Fall plantings, executed as perennials are going dormant, are less at risk of desiccation. Before the ground is frozen, they will root in even as top growth has ceased and nutrients are being drawn back into belowground storage parts. However, avoid planting small materials like plugs too late in the season or they may frost heave in the winter.

Mulching is an excellent practice to preserve moisture in well-drained soils. It provides water retention and several other benefits besides. A mulch of leaves or wood chips holds moisture in the soil, and feeds the soil biology around the newly introduced plant. A mulch application can suppress the inevitable germination of weeds in the newly disturbed soil created by the planting process as the shovel lifts, churns, and drops soil.

Live plant materials are available in a variety of forms, including container plants, plugs, and bare root plants. Balled and burlapped (B&B) materials are seldom practical for restoration use.

Container plants are available in pot sizes ranging from 3" cubes (more or less) to quarts on up to multiple gallon pots. Sometimes woody plants, especially trees used in landscaping, are available in truly large sizes. Be aware that these will need lots of TLC and water to transplant successfully. Often a smaller woody plant will establish better and may even catch up with a more mature transplant.

Container plants are more expensive than other options, but carry several advantages. They are available at almost any time during the growing year. Provided you have adequate water, you could pull off a great planting mid-July, not really an option for bare root material. They can have excellent survivorship and predictability.

Small container plants aggregated into a single press-formed tray with depressions are called **plugs** and are relatively economical. Often numbering 32, 38, 50, or even 72 plugs per tray, these plants are small but often poised to grow swiftly in new environments. For wild plantings, the larger and especially deeper plugs are more likely to possess adequate roots to survive. I find two-inch plugs too shallow to be worth the effort.

Bare root materials are usually available in a limited seasonal window, usually in early spring (sometimes in autumn as well). These are generally planted while still dormant, and will need to root in and leaf out more or less simultaneously. This can

Planting a plug tray of bottle gentian (*Gentiana clausa*).
CREDIT: RACHEL MACKOW

affect survivorship and make droughts even more risky. However, bare root materials

Area to be seeded.

Mixing the seeds with inert carrier, with Johann Rinkens.

are often inexpensive relative to other live plant materials. They are field grown and dug and don't come with the overhead expenses of potting soil and a container.

Seeding

Seeding can be more economical than live plantings, especially at large scales. Some species not readily available as live plants may be available as seeds. Then again, the converse is also true, and there are certain species (especially spring-flowering forest herbs) with recalcitrant seeds that don't survive drying or freezing, and need to be sown immediately after collecting. These are seldom practical to use unless you can collect them yourself.

Seeding can be done almost anytime other than mid-summer. However, a fall or winter seeding offers significant advantages in promoting germination the following spring. Freeze-thaw cycles and snowfall can help guarantee good seed-soil contact. Many native plants require a period of cold in order to germinate. Spring seedings may need to be worked into the soil more than fall seedings, through tillage and raking or drill seeding. Spring seedings may result in a heavier cover of grasses, which generally don't need winter stratification to germinate.

Areas of two acres or less can be easily seeded by hand. Seed should be "watered down" with an inert, lightweight grit medium (unscented cat litter, ProField, etc.)

Close-up of the seed mix.

Dividing the seed mix evenly.

and slightly moistened, then broadcast in wide sweeping arcs.

Mechanical, tractor-drawn drill seeders can also be effective, especially those which accommodate "fluffy" seed well. For large grassland seedings, equipment and assistance may be available through the Natural Resources Conservation Service in the United States.

Most seedings benefit from the addition of a non-persistent companion/cover crop for quick site cover. Oats is a good all-purpose cover, and winter rye, annual rye, and Canada wild rye can also be used in this capacity. The choice of cover crop will depend on the site, season of seeding, and one's ability to cut back the cover crop before it goes to seed or drowns out desirable seedlings.

Boneset (*Eupatorium perfoliatum*) seedling.

Blazing star (*Liatris spicata*) seedling.

Access

Projects with poor access are practically doomed from the start. If it is hard to reach your site for monitoring, it will be even harder to find the time to get equipment, plant materials, or fencing to a site. Scale-appropriate access is a must — foot trails, cart paths, and roads or parking at larger sites. True, roads and paths may degrade your site. I believe the appropriate response is to mitigate the damage rather than to ignore the necessity for access, except at the most pristine sites with the most committed personnel. On such sites, restoration practitioner Leslie Sauer recommends the use of "Franklin fire lines." Based on Ben Franklin's fire-fighting technique, where people passed buckets down the line instead of everyone running to and fro with their own bucket, this way of transporting materials minimizes trampling in sensitive habitats and confines damage to narrow routes.[412]

Post-Seeding Maintenance for Meadows

The general principle of post-seeding meadow maintenance is simple: to allow native seedlings to establish, while suppressing weeds. This is mostly accomplished through strategic mowing, though controlled burning is ideal in many areas if practicable.

In year one of the seeding (the first growing season after seeding occurs) it is recommended to mow the site every time vegetation becomes taller than ten inches. Mowing down to four inches, but not lower, is ideal. This keeps annual and other weeds from unduly shading native seedlings, and from going to seed in great quantity.

In practice, one to three mowings will often be necessary the first season. If Japanese stiltgrass is present as a management concern, time one mowing to occur around Labor Day (early September), just before stiltgrass seed set.

In the second year after seeding, annual and biennial weeds may still need to be controlled by mowing. Ideally, mowing would be timed to correspond with the flowering of prominent weed species.

Problematic perennial weeds are likely making themselves obvious by now and should be identified and removed before they become major management issues.

Mowing in the second year should be to a greater height, 8 to 12 inches if equipment allows. A flail mower is ideal, but a mechanical mower of any type, or even a scythe, can prove effective.

Meadow Maintenance

Once a seeded meadow is well established and native plants are dominating the site (ideally by year three), the area should be maintained with a once-annual mowing. This prevents volunteer woody plants from taking over the meadow.

The default mowing time is in the winter, when desirable plants have completed seed set and the ground is frozen (especially in wet sites). Other times of year may be appropriate given specific aesthetic or management objectives. For example, a May mowing may stymie Eurasian cool-season grasses, or a June mowing might maintain a slightly shorter meadow while allowing late-season flowers and grasses to bloom.

Perennial non-native weeds such as mugwort, Chinese bush clover, creeping thistle, and others should be monitored for and eliminated while occurrences are still small.

Conclusion: A Missing Link

THERE IS A LINK MISSING in our relationship to wild plants. It involves the consumption and management of wild plants for food, medicine, and cultural materials. This link traditionally connects us in deep reciprocity with the plant world. This connection founders at a time of deep alienation from nature, especially wild plants.

The conservation movement has served us well — humans and otherwise. This ethos, epitomized by "take only photographs, leave only footprints," has protected vast landscapes from direct exploitation and destruction. Walling off national parks and nature preserves from many forms of human utilization has been a necessary intervention, like putting a cast on an injured leg to prevent it from further injury.

But just as a cast is not a desirable long-term solution, neither should the recreation-only approach become our only paradigm for participating in nature. We need more from nature — we need meaning, community, sustenance. The other beings of the world, plants and animals, need more from us — understanding, restoration, emplacement. They need us to notice, for example, when the weather dramatically alters in the course of a few decades, and not to deny it from air-conditioned domiciles.

We have begun the process of healing by protecting some parts of the natural world. To recognize the whole Earth as nature, we'll need a deeper covenant than arises in an afternoon's hike in a local park.

Conservation has taken us remarkably far, and those who dedicated their lives to protecting great wild places, be they national parks or the nature preserve up the road, deserve much gratitude.

To truly restore nature at the necessary scale, we need to go beyond protecting wilderness areas, and include human habitats in a restoration process. Our yards, farms, and cities need to support nature in the same way that preserved areas do. As the authors of *Conservation Through Use: Lessons from the Mesoamerican Dry Forest* put it:

> There has been a growing recognition that it is not possible to create sufficient protected areas to conserve all species ... what happens outside protected areas will be as crucial for effective conservation of biodiversity as protection of in situ reserves. [413]

To continue the process of healing is to move from the cast (being walled off, immobilized) to an active and dynamic involvement. This is the restoration process. Without it, we will keep applying environmental bandages until the Earth is covered with them, but our underlying schism with the other living beings will remain.

Active and dynamic involvement is not hard to figure out. Models for humans acting as ecologically emplaced beings are everywhere, in our pasts and among the "others" that fall outside industrial civilization. Fortunately for us in North America, there is a rich heritage of Indigenous cultures that practiced — and still practice — sustenance, community, and meaning in sophisticated, locally relevant, and still-accessible ways. Traditional Ecological Knowledge can lead the way to much ecological repair. Those of us from outside of Indigenous cultures (like myself) need to find respectful ways to be supportive participants.

This book has been about one path towards sustenance, community, and meaning. A plant path. Plants are foundational to foodways and to the character of place, and also to healing. In this book, I've argued that plants are healers on several planes, at the level of the individual (through medicine and food), the community (through meaningful culture), and the place (through ecological restoration).

We may be on the cusp of trading in our very humanity, our animal selves, for the repeatable and quantifiable perfection of a technologized existence. We may be told that artificial intelligence is life, that life is whatever gene modification can make it, that we humans are but a transitional phase towards something more sophisticated, albeit dubiously alive. The biological and technological might merge.

To make the right decisions about what our humanity really is, to cast off the narrative that our existence is inherently evil or sinful, we need to be grounded in what it means to be human, to be animals, to be necessary participants in the ecological community.

Belonging to This Place

One of the first European-American scholars to take a deep interest in Indigenous peoples' uses of plants was Melvin Gilmore, an early ethnobotanist active in the first decades of the 20th century. Working with a variety of cultures of the Plains and upper Midwest, he documented plant uses for medicine, food, and craft, producing the book *Uses of Plants by the Indians of the Missouri River Region* as well as a series of monographs.[414]

He was among the first, in academia at least, to pay attention to Indigenous peoples' dispersal of plants, noting for example how sweetflag was found near many village sites.

Gilmore is a largely uncredited founder of the native plants movement, advocating throughout his career for adoption of indigenous plants and lifeways as a signal part of the new American nation. He wrote:

> The people of the European race in coming into the New World have not really sought to make friends of the native population, or to make adequate use of the plants or the animals indigenous to this continent, but rather to exterminate everything found here and to supplant it with the plants and animals to which they were accustomed at home.[415]

Gilmore did more than document Indigenous lifeways as museum pieces, or curiosities, or as a way to fit some "stage" of human development onto the timeline of "progress." I believe that Gilmore's painstaking fieldwork, notes, and writings point not just towards a vanishing past, but towards a future he hoped would be realized to the benefit of all:

> We shall make the best and most economical use of all our land when our population shall have become adjusted in habit to the natural conditions. The country can not be wholly made over and adjusted to a people of foreign habits and tastes. There are large tracts of land in America whose bounty is wasted because the plants which can be grown on them are not acceptable to our people. This is not because these plants are not in themselves useful and desirable, but because their valuable qualities are unknown.

Every culture represents a response to the human condition, a set of malleable answers to ecology and existence. What possibilities were foreclosed when Indigenous cultures were swept away by plagues and the colonial invasion? A critical set of ideas about sustenance, nature, and human craft that could do much to inform solutions to our current world crises.

Historical ecologists like William Denevan and Clark Erickson have rendered a sketch of Amazonian culture groups and civilizations that approached what we might call agriculture in a different, more complex way than the Eurasian norm. They describe a uniquely Amazonian cultivated landscape, where existing ecologies are optimized for abundance, diversity, and resiliency. We are realizing that the "virginal" Amazon, the greatest wilderness on the planet, was in fact born of a process in which humans are as fundamental as beavers are to a marsh.

Eastern North America was also mosaic of "wild" and cultivated landscapes, and meldings of the two were not well understood by European explorers or colonists.

In the Appalachian heartland, Cherokee "reshaped the land toward making bounty insurable," according to historian Tom Hatley. "In the Southern Appalachian highlands, where plant diversity and endemism are in the highest global rank of temperate zone sites," he writes, "agricultural diversity corellate[d] with native botanical diversity."[416] Across the shifting years, Cherokee tended orchards of native wood plants such as "persimmon, honey locust, Chickasaw plum, red mulberry, shellbark hickory, and black walnut,"[417] native cultigens like sunflower, sumpweed, maypops, maygrass, cresses, and chenopods,[418] as well as domesticates like squash, gourds, beans, and corn, with different Amerindian provenances. From a landscape perspective, this was a "system of cultivation cross-bedded into a broader ecosystem — a designed landscape."[419]

Florida's Timucua cultivated native tree and shrub crops such as "plums, persimmons, black cherries, mulberries ...

berries blue and raspberries of various kinds, grapes, nuts, walnuts, chestnuts, [and] dwarf chinquapin."[420] In addition to these upland species, around Lake Okeechobee they grew groundnut (*Apios americana*), as well as Southeast natives Florida arrowroot (*Zamia integrifolia*), and bullbrier (*Smilax bona-nox*),[421] both starchy root vegetables.

In Maine, Kerry Hardy, author of *Notes on a Lost Flute*, reports remnants of centuries-old stands of Jerusalem artichoke and bur oaks, both away from the center of their range farther west, as well as fire-maintained colonies of native lilies and American plum, and "oak openings" similar to the savanna landscapes of the Midwest. These traces epitomize the land management that the Algonquin termed *minisage*: to tend the land, to burn the underbrush. Plants that Hardy lists as beneficiaries of the fire-maintained landscape include "a veritable Indian grocery store list: strawberries, raspberries, blueberries, chokecherries, plums, groundnut, wood lily, Canada lily, Jerusalem artichoke, hickories, bur oaks, and butternuts ... plants that fed not only Indians but also many birds and mammals, themselves in the Wabanaki food chain."[422]

Not all Indigenous land management lies in the past. Where Indigenous peoples have sovereignty over significant areas of land, the influence of cultural values and Traditional Ecological Knowledge can still lead to very different environmental outcomes than in similar, neighboring landscapes managed by non-Indigenous people. A 2018 study by Donald Waller and Nicholas Reo compared the diversity, composition, and structure of Menominee and Ojibwe Nation forests in northern Wisconsin to nearby county, state, and national forests. While most forest sites in the area declined in native plant diversity since the 1950s, small increases in diversity were found in the two Ojibwe reservations. Northern white cedar and hemlock recruitment was up to 5.65 times greater than in national forests, reflecting the mature forest conditions found on Indigenous lands, including "rotting logs and cool, moist microsites," as well as significantly lower deer densities.[423]

The Menominee and Ojibwe "manage deer assertively," with no hunting season or harvest limits. Deer hunting rules are "enforced socially, not by wardens. Ojibwe hunters seek venison to support community needs, not individual interests." Indigenous respect for wolves and other predators may also play a role in the better forest health of Indigenous lands. "The Ojibwe, in particular, revere and respect wolves (*ma'iingan*), considering them to be their brothers whose fate is tied directly to that of their Indigenous Nation. Ojibwe hunters view ma'iingan not as competitors but rather as family members who, like people, need to eat. To reinforce this

kinship, they share their deer with wolves, eagles, and bears by leaving ceremonial offerings and unused portions of animals in the woods."[424]

As a result of these cultural practices, forest land stewarded by the Ojibwe and Menominee have better ecological health (as indicated by biomass, carbon storage, and plant diversity), as well as better forest yields and natural regeneration than neighboring government-managed land, despite those public lands' stated goals of protecting ecological conditions, soils, and wildlife populations.[425]

If we, the products of so many far-flung diasporas, are to belong to this North American land, we might follow Melvin Gilmore's advice to appreciate the plants of our new home and the bounty they represent. We might also realize that this work was already done in the past, in a way that offers brilliant promise to the problems of monoculture, disconnection from nature, and environmental collapse which plague our present existence.

There is a deep tension in our society between wild and agricultural landscapes, between environment and economy. Perhaps part of the solution to these tensions was already innovated here in the Americas, centuries in the past, by making these oppositional forces one and the same. By treating biodiversity as food diversity, and farming as tending. Ecological restoration becomes less about removing human impacts and more about directing human activity in ways that most benefit the whole ecology of a place ... including us.

The beaver offers itself as a totem of sorts, an archetype of an ecosystem engineer that benefits the environment and itself. This fiercely meddlesome creature ravages trees, clogs streams with vast detritus, converts uplands into flooded aquaculture areas and watery transportation hubs. In pre-colonial times, beavers dammed and channeled as they went, spreading prolifically across the landscape until every landscape was permeated with them. In the process, beavers created habitats that support vast numbers of plant and wildlife species, which are left bereft without their activity. Beavers force the infiltration of surface water into groundwater through the sheer weight of their ponds. They slow the rush of floodwaters through a filtery matrix of biomass, retaining prodigious amounts of fertile soil in the beaver meadows they leave in the wake of their ever-shifting creative destruction.[426]

We pat ourselves on the back for our smarts, our tools, for the written word, for civilization and technology. Can we be as brilliant as these toothy rodents, working immersed in the promise of our bioregion? If so, we will do so by continuing a story long told by Indigenous cultures in the New World, using ecological techniques inspired by the Wabanaki, Cherokee, Timucua, Menominee, Ojibwe, and others, and finding belonging in this place.

Paper birch.

Appendix – Assessment and Monitoring Techniques

Assessment and Monitoring

IN ORDER TO MAKE objective decisions about sites, we need to look at methods for habitat assessment and monitoring.

Monitoring often gets left out of actual practice, by time- and money-constrained practitioners, by those who fly by the seat of their pants, and other creative types.

Monitoring can take many forms, including practices that will appeal to the budget conscious, and even to creative types (guilty as charged). The goal of monitoring, however, remains constant and always worthwhile. Monitoring is a practice that allows us to perceive change over time and to learn from our successes and failures. Many of us consider our projects to be models for others to learn from. Without proper monitoring and assessment, you're not really creating a model that others can learn from or imitate.

Baseline Monitoring

In baseline monitoring, we generate information about the site as it is, prior to any management activities. We develop a complex picture of the life of the place, including the living beings that presently may inhabit it, as well as the qualities of the abiotic features present.

Baseline monitoring serves two primary purposes. It establishes a "before" against which to contrast later monitoring, post-management. Baselines also inform our understanding of place so that appropriate decisions can be made pertaining to site preparation, plant materials selection, and working with animal and fungal partners.

Monitoring Methods

Photo Monitoring

Photography can be used as a monitoring tool by establishing static locations and angles from which to take sets of photos on an annual (or other recurring) basis. This type of monitoring has a lot of communicative strength (consider the "before" and "after" shot) and is effective at capturing general trajectories. Properly framed, structural changes may be better expressed in this medium than in any of the other monitoring methods below. Photographs may also record information we did not think to gather, making it available for interpretation once its importance is realized.

Biological Inventories

A mainstay of monitoring involves the biological inventory. This often takes the form of a species list. The location the list is taken in should be circumscribed in space (using quadrats, transects, plots, sections, and/or stands) and inventory work should recur on a regular basis, such as every few years.

Some of the species groups we may choose to track include plants, birds, pollinators, amphibians — each may communicate richly about some specific attribute of our restored habitats, and each may interrelate to create a robust picture when combined.

One of the pleasures of monitoring is the natural flow of learning, as new species and groups become familiar and enrich our perceptions. Good field guides, manuals, and identification tools are key in determining the identity of unfamiliar species.

One of my favorite biological inventory techniques is Floristic Quality Assessment, because of the way that it combines deep attention to plant species with quantitative, actionable metrics.

Floristic Quality Assessment

Let's say you walk into my office with two plant specimens and dump them on my desk with a challenge. "OK, Mr. Botanist, tell me about the places where these plants came from." One specimen is pokeweed, the other a showy orchis.

The showy orchis (*Galearis spectabilis*) is easier. It probably came from a rich moist mature forest with diabase, limestone, or other basic rock type influencing the soil chemistry. In my area, it is frequently associated with species like tuliptree, spicebush, and black cohosh.

The pokeweed would be tough. It could be from the edge of a vegetable garden or a roadside — or a tip-up mound in an old growth forest (or logging road when that forest is cut as "overmature" timber). This species responds to soil disturbance and light availability pretty much wherever it happens, and does well in a rich soil. It is a weedy and opportunistic species that does not require an intact native habitat.

The differing character of these two plant species can be described as a difference in "quality." Just such a concept underlies the method of describing plant communities called Floristic Quality Assessment (FQA). Floristic Quality Assessment is a method for describing the quality and diversity of wild and restored plant communities. It translates plant inventories into metrics that can be used to compare between sites or across time.

FQA is based on two fundamental tenets:

1. Some plant species exhibit fidelity to particular native habitats whereas others do not.
2. Plant species have differing tolerance for human disturbance.

Plant species with a high fidelity to intact native habitats are called "conservative" species. Plants on the opposite end of the spectrum are those we might term "weedy," "opportunistic," or "ruderal." Habitats with high numbers of conservative species are described as high quality habitats.

In FQA, each plant species in a state's (or bioregion's) flora is assigned a number from 0–10 reflecting its quality. These numbers are known as Coefficients of Conservatism (no, it's not a party of right-wing mathematicians) or sometimes the "C value."

Pokeweed properly deserves a C value of 1 (zeroes are for non-native species), and showy orchis's C value should be somewhere closer to a 10. Not only does the orchid require a relatively undisturbed habitat with specific geology and moisture, it (like other terrestrial orchids) can only recruit at a site if certain soil fungi are present. Its needs are specific but it can be a very long-lived and resilient species in the proper niche. By way of contrast, the pokeweed might come on strong in the same habitat (given some disturbance) but there's no way it could persist as long, be as tolerant of shade and other stressors, or build as enduring an architecture as the orchid.

All plant species will be somewhere on the spectrum between 0 and 10, some more generalist, others more conservative.

Spicebush seems to me to exemplify a perfect 5 (though it is not necessarily exactly a 5 on all state lists). It exhibits fidelity to a real native habitat: rich moist forest (maybe even the same habitat as the orchid). So its C value should be somewhere in the range of 5–10. However, it is easily dispersed (primarily by birds), can recruit readily into post-agricultural woodlands, is very tolerant of logging and other canopy disturbances, and is generally a colonial, tough customer. Given its fidelity for rich forest (including some of the highest quality habitats in my area) it can hardly be termed a weed. But given its easy dispersal and recruitment and high tolerance for disturbance, it isn't a very conservative species either. It sits comfortably in the middle position.

This all translates into metrics reflecting a site's quality in the following way. An inventory containing a list of plants found on the site (or in the plots, transects, etc.) is uploaded to a database. I presently use the online database at UniversalFQA.org. Here, each plant on your inventory is matched with the C value from the pertinent state or regional flora where your inventory took place. C values have been assigned by groups of local botanists for most of the United States.

Table 1: FQA Metrics

Term	Description
Total species richness	Total number of native and non-native species.
Native species richness	Number of native species.
Total mean C	Mean conservatism coefficient for all native and non-native species.
Native mean C	Mean conservatism coefficient for native species.
Total FQI	Floristic quality index: total mean C multiplied by the square root of the total species richness.
Native FQI	Floristic quality index: native mean C multiplied by the square root of the native species richness.
Adjusted FQI	Adjusted floristic quality index: 100 multiplied by the native mean C divided by 10 and multiplied by the square root of the native species richness divided by total species richness.

Adapted from UniversalFQA.org

Once the entire inventory has a C value assigned, this can be printed or viewed in a variety of ways. Further, a number of metrics are automatically generated by various manipulations of the C values. Table 1 shows some of the metrics available from UniversalFQA.org.

By manipulating the values of all the plants on an inventory, metrics can be derived that suggest the character, diversity, and naturalness of a restoration site or a wild area.

I tend to use the Native Mean C as a simple metric to get a handle on the quality of a site. It's a simple average of the C values of all the native plants in the inventory. Even if the site is beat up, invaded and deer browsed, a high Native Mean C suggests to me that a site was once high quality,

contains relict specialist species, and may have high restoration potential.

For more precise work, I often consider the Total FQI. This works best when comparing survey units of fixed size, such as plots or quadrats. Because one factor in the calculation of this metric is the species richness, it's a pretty useless metric to compare between sites of dramatically unequal sizes. A middling two hundred-acre site will have higher species counts than a really nice five-acre fragment.

Below is an example (see Table 2) of the metrics associated with a 100m² plot I inventoried on a high quality dolomite cliff I did botanical survey work at in 2017. The Native Mean C is extremely high at 6.4. Even a brief scan of the C values associated with each species shows that most of them are over 5.

Table 2: Example of the Metrics with a 100m² Plot

Plot 3			
Conservatism-Based Metrics:			
Total Mean C:	5.8		
Native Mean C:	6.4		
Total FQI:	33.3		
Native FQI:	35.1		
Adjusted FQI:	61		
% C value 0:	9.1		
% C value 1–3:	6.1		
% C value 4–6:	42.4		
% C value 7–10:	42.4		
Native Tree Mean C:	6		
Native Shrub Mean C:	6.5		
Native Herbaceous Mean C:	6.5		
		% of Total	
Species Richness:			
Total Species:	33		
Native Species:	30	90.90%	
Non-native Species:	3	9.10%	
Species Wetness:			
Mean Wetness:	2.7		
Native Mean Wetness:	2.6		
Physiognomy Metrics:			
Tree:	8	24.20%	
Shrub:	4	12.10%	
Vine:	1	3%	
Forb:	13	39.40%	
Grass:	1	3%	

Table 2: Example of the Metrics with a 100m² Plot (*Continued*)

Plot 3			
		% of Total	
Sedge:	3	9.10%	
Rush:	0	0%	
Fern:	3	9.10%	
Bryophyte:	0	0%	
Duration Metrics:			
Annual:	0	0%	
Perennial:	33	100%	
Biennial:	0	0%	
Native Annual:	0	0%	
Native Perennial:	30	90.90%	
Native Biennial:	0	0%	
Species:			
Scientific Name	**Native?**	**C**	**Common Name**
Acer saccharum	native	5	sugar maple
Amelanchier arborea	native	6	serviceberry
Aquilegia canadensis	native	8	wild columbine
Aralia nudicaulis	native	5	wild sarsaparilla
Asarum canadense	native	5	Canadian wild ginger
Asplenium trichomanes	native	8	maidenhair spleenwort
Carex eburnea	native	10	ebony sedge
Carex laxiflora	native	5	loose-flowered sedge
Carex pedunculata	native	9	long-stalked sedge
Cornus rugosa	native	7	round-leaved dogwood
Dryopteris marginalis	native	7	marginal wood fern
Euonymus alatus	non-native	0	winged euonymous
Eurybia divaricata	native	5	white wood aster

Table 2: Example of the Metrics with a 100m² Plot (*Continued*)

Plot 3			
Scientific Name	**Native?**	**C**	**Common Name**
Fagus grandifolia	native	6	American beech
Fraxinus americana	native	5	white ash
Hamamelis virginiana	native	6	witch hazel
Hepatica nobilis	native	7	roundlobe hepatica
Lonicera maackii	non-native	0	amur honeysuckle
Maianthemum racemosum	native	5	false Solomon's seal
Mitchella repens	native	5	partridgeberry
Packera obovata	native	8	roundleaf ragwort
Parthenocissus quinquefolia	native	2	Virginia creeper
Patis racemosa	native	7	blackseed ricegrass
Polygonatum pubescens	native	6	Solomon's seal
Polypodium virginianum	native	8	polypody
Prunus serotina	native	2	wild black cherry
Solidago caesia	native	6	bluestem goldenrod
Taxus canadensis	native	10	Canadian yew
Thalictrum dioicum	native	7	early meadow-rue
Tsuga canadensis	native	8	eastern hemlock
Veratrum hybridum	native	7	slender bunchflower
Veronica officinalis	non-native	0	speedwell
Viburnum acerifolium	native	6	maple-leaved viburnum

The FQA methodology was originally developed by Gerould Wilhelm and Floyd Swink in the Midwest. It was a means to positively identify remnant prairie sites in a region where over 99% of the original prairie had been destroyed. Subsequent work pioneered its applicability for ecological restoration as well.

Sampling Units

In order to gain statistical strength, inventory work often utilizes replicate sampling units of some type. These are circumscribed areas that can be analyzed as a group to add fidelity to a picture and to answer questions about abundance and diversity.

Typical sampling units include quadrats, which are typically small sample areas (such as 1m² or smaller). Plots are larger and can provide a truer picture of composition of macrofeatures such as trees. I frequently use 100m² plots in my survey work. This may sound large, but it's actually a modest circular plot with a radius of about 5.6 meters. I put a temporary stake in the ground in the center of the plot, attach a string of the desired radius, and survey within the circular confines of the area delineated by the string.

Transects are linear survey paths that often have either quadrats or plots placed at designated intervals along the line. For example, one quadrat could be placed every 10 meters.

Locations of sampling units can be randomized in order to limit subjective biases that might influence a practitioner to place plots in specific kinds of places (easy places to access, nice places, etc.) Randomization can be technologically sophisticated, based on Geographic Information System (GIS) map points and random number generators, or can be simpler, such as randomizing intervals between quadrats placed on a transect line.

Various modifications have been built in to FQA to include species abundance of percent cover. My preference at this point is to use replicate quadrats or plots to assess frequency and changes in metrics. For example, if Japanese barberry is found in eight out of ten plots before restoration, and white wood aster is found in two out of ten, and then after restoration barberry is found in only one out of ten, and white wood aster in six out of ten, we have solid quantitative information to measure success with. Specific FQA-related changes that can be assessed using replicate quadrats include shifts in Total Mean C, or in the percentage of plots with conservative species (C value 7–10). Or, the Total FQI of each plot can be averaged and compared with prerestoration baseline metrics.

Endnotes

Introduction

1. Samuel Thayer, *Incredible Wild Edibles: 36 Plants That Can Change Your Life,* (Bruce, WI: Forager's Harvest, 2017), 432. Thayer writes: "Unlike agriculture, ecoculture considers both ecology and production."

Chapter 1

2. Kevin C. Ryan, Eric E. Knapp, and J. Morgan Varner. "Prescribed fire in North American forests and woodlands: history, current practice, and challenges." *Frontiers in Ecology and the Environment* 11, no. s1 (2013): e15–e24.

3. William M. Denevan, "The pristine myth: the landscape of the Americas in 1492," *Annals of the Association of American Geographers* 82, no. 3 (1992): 369–385.

4. Eric W. Sanderson, *Mannahatta: A Natural History of New York City,* (Abrams, 2013), 10.

5. Ibid., 126

6. Arthur Haines, *A New Path,* (V.F. Thomas Co., 2017), 63.

7. Fikret Berkes, "Traditional Ecological Knowledge in Perspective," *Traditional Ecological Knowledge Concepts And Cases* (IDRC, 1993), 11.

8. Kimmerer and Lake, "Maintaining the Mosaic," 36–41.

9. Robin Wall Kimmerer and Frank Kanawha Lake, " Maintaining the Mosaic: The role of indigenous burning in land management." *Journal of Forestry* 99, no. 11 (2001), 36.

Chapter 2

10. Elizabeth Bach, "Soils," interview with Matt Candeias, *In Defense of Plants Podcast,* Episode 35, November 11, 2015, audio. www.indefenseofplants.com/podcast/2015/11/15/ep-35-soils

11. J. Schwartz, "Soil as Carbon Storehouse: New Weapon in Climate Fight?" March 4, 2014. e360.yale.edu/feature/soil_as_carbon_storehouse_new_weapon_in_climate_fight/2744/

12. Illima Loomis, "Trees in the Amazon make their own rain: Scientists uncover why it starts raining in the region several months before it should," *Science,* August 4 2017. www.sciencemag.org/news/2017/08/trees-amazon-make-their-own-rain

13. E. L. Stone and P. J. Kalisz. "On the Maximum Extent of Tree Roots," *Forest Ecology and Management,* vol. 46, (no. 1, 1991), 59–102.

14. Arthur Haines, *Ancestral Plants: A Primitive Skills Guide to Important Edible, Medicinal, and Useful Plants Volume 2,* (Anaskamin, 2015), 13.

15. Dhiraj A. Vattem, K. Mihalik, Sylvia H. Crixell, Robert McLean, "Dietary phytochemicals as quorum sensing inhibitors," *Fitoterapia* 78, no. 4 (2007): 302–310.

Chapter 3

16. Simon Thakur, "Mirror Neurons, Shape Shifting & the Body Map," interview with Daniel Vitalis, *Rewild Yourself,* Episode 133, March 22, 2017. www.danielvitalis.com/rewild-yourself-podcast/mirror-neurons-shapeshifting-the-body-map-simon-thakur-133

17. Tyler Smith, Kimberly Kawa, Veronica Eckl, Claire Morton, Ryan Stredney, "Herbal supplement sales in US increased 8.5% in 2017, topping $8 billion," *HerbalGram* 119 (2018), 66.

Chapter 4

18. D. Jaffe, "The Fluttering Forest: Native Trees and Shrubs Support the 'Other Pollinators,'" *Native Plant News*, New England Wildflower Society, Fall-Winter 2018.

Section 2 (Introduction)

19. Friend and colleague Johann Rinkens of Field Without Fences worked with me on this project.

Chapter 5

20. Michael Kudish, *Adirondack Upland Flora: an Ecological Perspective,* (Chauncy Press, 1992), 55–56.

21. John A. Conners, *Groundwater for the 21st Century,* (McDonald & Woodward Pub. Co., 2012), 67–69.

Chapter 6

22. M. Kat Anderson, *Tending the Wild,* (University of California Press, 2005).

23. Robin Wall Kimmerer, *Braiding Sweetgrass: Indigenous Wisdom, Scientific Knowledge and the Teachings of Plants,* (Milkweed Editions, 2013), 163.

24. Ibid., 165.

Chapter 7

25. Chuck Stead, "Ramapough/Ford: The Impact and Survival of an Indigenous Community in the Shadow of Ford Motor Company's Toxic Legacy," Dissertations & Theses (2005). Accessed January 14, 2022 at aura.antioch.edu/etds/200

26. Jonathan T. Bauer, Liz Koziol, James D. Bever, "Ecology of Floristic Quality Assessment: testing for correlations between coefficients of conservatism, species traits and mycorrhizal responsiveness," *AoB Plants* 10, no. 1 (2018), 9.

27. Charles D. Canham and Lora Murphy, "The demography of tree species response to climate: sapling and canopy tree survival," *Ecosphere* 8 (2) (February 2017), 9.

28. James Baird, *The ecology of the Watchung Reservation, Union County, New Jersey; a description of the biotic communities and recommendations for their management prepared at the request of the Union County Park Commission,* (New Brunswick: Dept. of Botany, Rutgers, the State University, 1956).

Chapter 8

29. Timothy P. Spira, *Wildflowers and Plant Communities of the Southern Appalachian Mountains and Piedmont: A Naturalist's Guide to the Carolinas, Virginia, Tennessee, and Georgia,* (University of North Carolina Press, 2011), 192–193.

30. Roger Earl Latham, John E. Thompson, Sarah A. Riley, Anne W. Wibiralske, "The Pocono till barrens: shrub savanna persisting on soils favoring forest," *Bulletin of the Torrey Botanical Club* (1996), 343.

31. Spira, *Wildflowers and Plant Communities,* 205.

32. Roger Earl Latham and J. F. Thorne, "Keystone Grasslands: Restoration and Reclamation of Native Grasslands, Meadows, and Savannas in Pennsylvania State Parks and State Game Lands," (Harrisburg: Wild Resource Conservation Program, Office of Conservation Science PA DCNR 2007), 11.

33. Ben Goldfarb, *Eager: The Surprising, Secret Life Of Beavers And Why They Matter,* (Chelsea Green Publishing, 2018).

34. William Moomaw, Gillian Davies, Max Finlayson, "What the world needs now to fight climate change: More swamps," *The Conversation* (September 12, 2018), theconversation.com/what-the-world-needs-now-to-fight-climate-change-more-swamps-99198

35. Conners, *Groundwater for the 21st Century*, 28.

36. Louise Wootton, Jon Miller, Christoper Miller, Michael Peek, Amy Williams, Peter Rowe, *Dune Manual,* (New Jersey Sea Grant Consortium, 2016), 27–28.

37. Wootton, et al., *Dune Manual*, 31.

38. R. Alan Shadow, "Plant fact sheet for Sea Oats Uniola paniculata L." (USDA-Natural Resources Conservation Service, East Texas Plant Material Center, October, 2007), 1.

Chapter 9

39. Héloïse Coté, Marie-Anne Boucher, André Pichette, Benoit Roger, Jean Legault, "New antibacterial hydrophobic assay reveals Abies balsamea oleoresin activity against Staphylococcus aureus and MRSA," *Journal of Ethnopharmacology,* Volume 194 (2016) 684–689.

40. Abir Nachar, Ammar Saleem, John T. Arnason, Pierre S. Haddad, "Regulation of liver cell glucose homeostasis by dehydroabietic acid, abietic acid and squalene isolated from balsam fir (Abies balsamea (L.) Mill.) a plant of the Eastern James Bay Cree traditional pharmacopeia," *Phytochemistry*, Volume 117 (2015), 373.

41. Max Paschall, "Christmas Tree Farms & Climate Change: A Permaculture Perspective," Shelterwood Plants Blog (May 22, 2019). www.shelterwoodplants.com/blog/2019/5/22/christmas-tree-farms-and-climate-change-a-permaculture-perspective

42. Maine Natural Areas Program, "Red Maple Fen," accessed January 12, 2022 at www.maine.gov/dacf/mnap/features/communities/redmaplewoodedfen.htm

43. Andrew Conger, "A Comparative Analysis of Sugar Concentrations in Various Maple Species on the St. Johns Campus," accessed November 11, 2019 at citeseerx.ist.psu.edu/viewdoc/download?doi=10.1.1.583.3123&rep=rep1&type=pdf

44. Darryl Patton, *Mountain Medicine: The Herbal Remedies of Tommie Bass,* (Natural Reader Press, 2004), 171.

45. W. A. Geyer, J. Dickerson, J. M. Row, "Plant Guide for Silver Maple (Acer saccharinum L.)." (USDA-Natural Resources Conservation Service, Manhattan Plant Materials Center, November 2010). www.nrcs.usda.gov/Internet/FSE_PLANTMATERIALS/publications/kspmcpg10099.pdf

46. Alan S. Weakley, *Flora of the Southeastern United States,* (University of North Carolina at Chapel Hill Herbarium, October 20, 2020), 1435.

47. "Achillea millefolium," Native Plant Trust, gobotany.newenglandwild.org/ species/achillea/millefolium/ 8/5/17

48. Kelly Kindscher, *Medicinal Wild Plants of the Prairie,* (University Press of Kansas, 1992), 17–20.

49. Timothy J. Motley, "The ethnobotany of sweet flag, Acorus calamus (Araceae)," *Economic Botany* 48, no. 4 (1994), 401.

50. Kindscher, *Medicinal Wild Plants of the Prairie,* 25–26.

51. Sirius J, "What to Do if You Get Too High", *High Times,* hightimes.com/culture/what-to-do-if-you-get-too-high/

52. Francesca Borrelli and Edzard Ernst, "Cimicifuga racemosa: a systematic review of its clinical efficacy," *European Journal Of Clinical Pharmacology* 58, no. 4 (2002), 235–241.

53. Hans-Heinrich Henneicke-von Zepelin, "60 years of Cimicifuga racemosa medicinal products." *Wiener Medizinische Wochenschrift* 167, no. 7-8 (2017), 147–159.

54. American Herbal Products Association, "Tonnage Surveys of Select North American Wild-Harvested Plants, 2006–2010," (Silver Spring, Maryland: American Herbal Products Association, 2012), 5.

55. Henneicke-von Zepelin, "60 years of Cimicifuga racemosa medicinal products," 148.

56. Kelly Kindscher, *Edible Wild Plants of the Prairie*, (Lawrence, Kansas: University Press of Kansas, 1987), 15.

57. Richard Dwight Porcher and Douglas Alan Rayner, *A Guide to the Wildflowers of South Carolina*, (Columbia: University of South Carolina Press, 2001), 228.

58. Kindscher, *Medicinal Wild Plants of the Prairie*, 29–30.

59. Nancy Jean Turner and Patrick von Aderkas, "Sustained by First Nations: European newcomers use of Indigenous plant foods in tempterate North America," *Acta Societatis Botanicorum Poloniae* 81, no. 4 (2012), 302.

60. "Amelanchier bartramiana," *Native Plant Trust*, gobotany.newenglandwild.org/species/amelanchier/bartramiana/

61. Kindscher, *Edible Wild Plants of the Prairie*, 39–40.

62. Matthew Alfs, *Edible & Medicinal Wild Plants of the Midwest*, (New Brighton, Minnesota: Old Theology Book House, 2013), 187–198.

63. Peter Kalm, *Travels into North America: containing its natural history, and a circumstantial account of its plantations and agriculture in general, with the civil, ecclesiastical and commercial state of the country, the manners of the inhabitants, and several curious and important remarks on various subjects.* Vol. 1., 1770. content.wisconsinhistory.org/digital/collection/aj/id/14175

64. Kerry Hardy, email to the author, December 16, 2021.

65. Matthew Wood, *The Earthwise Herbal: A Complete Guide to Old World Medicinal Plants*, (Berkeley, California: North Atlantic Books, 2008), 92.

66. Kindscher, *Edible Wild Plants of the Prairie*, 51.

67. William Emery Doolittle, *Cultivated Landscapes of Native North America*, (Oxford University Press, 2000).

68. Anne Bruneau and Gregory Anderson, "Reproductive Biology of Diploid and Triploid Apios americana (Leguminosae)," *American Journal of Botany*. 75, (1988) 1876–1883.

69. Kindscher, *Medicinal Wild Plants of the Prairie*, 41–45.

70. Huron H. Smith, "Ethnobotany of the Forest Potawatomi Indians," Bulletin Of The Public Museum Of The City Of Milwaukee, Vol. 7, No. 1, (May 9, 1933), 40.

71. "Aralia nudicaulis," *Plants for a Future*, accessed November 3, 2020 at pfaf.org/user/Plant.aspx?LatinName=Aralia+nudicaulis

72. Smith published four manuscripts — in 1923, 1928, 1932, and 1933. The Ho-Chunk manuscript was revived and completed by Kelly Kindscher and Dana P. Hurlburt in 1998. The published manuscripts, his field notes, and herbarium specimens can be found in an excellent online exhibit of the Milwaukee Museum: www.mpm.edu/research-collections/botany/ online-collections-research/ethnobotany. The Ho-Chunk manuscript is published as Kelly Kindscher and Dana P. Hurlburt, "Huron Smith's ethnobotany of the Hocąk (Winnebago)," *Economic Botany* 52, no. 4 (1998), 352–372.

73. Kelly Kindscher, Loren Yellow Bird, Michael Yellow Bird, Logan Sutton, "Sahnish (Arikara) Ethnobotany," (2018 prepress copy courtesy of Kelly Kindscher), 43.

74. Tomislav Jurendić and Mario Ščetar, "Aronia melanocarpa: Products and By-Products for Health and Nutrition: A Review," *Antioxidants* 10, no. 7 (2021), 1.

75. Maria Handeland, Nils Grude, Torfinn Torp, Rune Slimestad, "Black chokeberry juice

(Aronia melanocarpa) reduces incidences of urinary tract infection among nursing home residents in the long term — a pilot study," *Nutrition Research* 34, no. 6 (2014): 518–525.

76. Jurendić and Ščetar, "Aronia melanocarpa," 11.

77. Doolittle, *Cultivated Landscapes of Native North America*, 42.

78. Matt Candeias, "North America's Native Bamboos," In Defense of Plants, www.indefenseofplants.com/blog/2017/6/26/north-americas-native-bamboos

79. Daniel E. Moerman, *Native American Medicinal Plants: An Ethnobotanical Dictionary*, (Timber Press), 2009, 96–97.

80. Erin Piorier, "Wild Ginger: A Love Song to a Toxic Plant," *Minnesota Herbalist*, minnesotaherbalist.com/2015/09/02/wild-ginger-a-love-song-to-a-toxic-plant/

81. Kelly Kindscher, Loren Yellow Bird, Michael Yellow Bird, and Logan Sutton, *Sahnish (Arikara) Ethnobotany*, Society of Ethnobotany, 49.

82. Alfs, *Edible & Medicinal Wild Plants*, 75–77.

83. Hannah A. Boone, Damir Medunjanin, Adna Sijerčić, "Review on potential of phytotherapeutics in fight against COVID-19," *International Journal of Innovative Science and Research* 5 (2020), 481–491.

84. Spira, *Wildflowers and Plant Communities of the Southern Appalachian Mountains and Piedmont*, 274.

85. Haines, *Ancestral Plants, Volume 2*, 56.

86. Spira, *Wildflowers and Plant Communities*, 227.

87. Brett J. Murphy, Richard E. Carlson, John D. Howa, Tyler M. Wilson, R. Michael Buch, "Determining the authenticity of methyl salicylate in Gaultheria procumbens L. and Betula lenta L. essential oils using isotope ratio mass spectrometry," *Journal of Essential Oil Research* (2021), 1–10.

88. Ronald J. Uchytil, "Betula papyrifera," in: Fire Effects Information System, [Online]. U.S. Department of Agriculture, Forest Service, Rocky Mountain Research Station, Fire Sciences Laboratory (Producer), accessed December 31, 2021 at www.fs.fed.us/database/feis/plants/tree/betpap/all.html.

89. Ibid.

90. Wootton, et al, *Dune Manual*, 24–25.

91. Arthur Haines, *Ancestral Plants: A Primitive Skills Guide to Important Edible, Medicinal, and Useful Plants Volume 1*, (Anaskamin, 2010), 88.

92. Michael Kudish, *Adirondack Upland Flora: An Ecological Perspective*, (Chauncy Press, 1992), 108.

93. Marc D. Abrams and Gregory J. Nowacki, "Native Americans as Active and Passive Promoters of Mast and Fruit Trees in the Eastern USA," *The Holocene* 18, no. 7 (2008), 1129

94. D. A. Tirmenstein, "Carya ovata," in: Fire Effects Information System, U.S. Department of Agriculture, Forest Service, Rocky Mountain Research Station, Fire Sciences Laboratory (Producer), accessed 1/11/2021 at www.fs.fed.us/database/feis/plants/tree/carova/all.html

95. Sarah H. Hill, *Weaving New Worlds: Southeastern Cherokee Women and Their Basketry*, (University of North Carolina Press, 1997), 9.

96. "Bitternut hickory," *Keystone Crops*, accessed January 3, 2022 at keystonetreecrops.com/news/bitternut-hickory

97. Hill, *Weaving New Worlds*, 9.

98. Douglass F. Jacobs, Harmony J. Dalgleish, C. Dana Nelson, "Synthesis of American chestnut (Castanea dentata) biological, ecological, and genetic attributes with application to forest restoration," Forest Health Initiative, accessed December 22, 2021 at foresthealthinitiative. org/resources/chestnutdossier. pdf

99. Doolittle, *Cultivated Landscapes of Native North America*, 67.

100. Janet Sullivan, "Castanea pumila," in: Fire Effects Information System, [Online]. U.S. Department of Agriculture, Forest Service, Rocky Mountain Research Station, Fire Sciences Laboratory, accessed January 13, 2022 at www.fs.fed.us/database/feis/plants/tree/caspum/all.html

101. Moerman, *Native American Medicinal Plants,* 124–125.

102. Richo Cech, "Blue Cohosh," in Rosemary Gladstar and Pamela Hirsch, eds. *Planting the Future: Saving our Medicinal Herbs,* (Inner Traditions/Bear & Co, 2000), 80–81.

103. David A. Perry, Ram Oren, and Stephen C. Hart, *Forest Ecosystems,* (Baltimore: JHU Press, 2008), 123.

104. Sara Leckie, Mark Vellend, Graham Bell, Marcia J. Waterway, Martin J. Lechowicz, "The seed bank in an old-growth, temperate deciduous forest," *Canadian Journal of Botany* 78, no. 2 (2000), 181–192.

105. Milo Coladonato, "Ceanothus americanus," in: Fire Effects Information System, [Online]. U.S. Department of Agriculture, Forest Service, Rocky Mountain Research Station, Fire Sciences Laboratory, accessed January 13, 2022 at www.fs.fed.us/database/feis/ [2016, November 26].

106. Kindscher, *Edible Wild Plants of the Prairie,* 77.

107. Stephen H. Buhner, *Herbal Antivirals: Natural Remedies for Emerging and Resistant Viral Infections,* (Storey Publishing, 2013), 257–259.

108. Wood, *Earthwise Herbal,* 107.

109. Buhner, *Herbal Antivirals,* 258.

110. Daniel E. Moerman, *Native American Food Plants: An Ethnobotanical Dictionary,* (Timber Press, 2010), 78.

111. Samuel Thayer, Nature's Garden: A Guide to Identifying, Harvesting, and Preparing Edible *Wild Plants,* (Forager's Harvest, 2010), 125–127.

112. Margaret B. Gargiullo, *A Guide to the Native Plants of the New York City Region,* (Rivergate Books, 2010), 7.

113. Green Deane, "Eastern Red Bud: Pea Pods on a Tree," Eat The Weeds, accessed January 13, 2022 at www.eattheweeds.com/eastern-red-bud-pea-pods-on-a-tree/

114. David Hoffman, *Holistic Herbal: A Safe and Practical Guide to Making and Using Herbal Remedies,* (Thorsons, 1990), 177.

115. David Winston, "Nvwoti: Cherokee Medicine," *Journal of the American Herbalists Guild,* 4.

116. "Iridoid Glycosides," Science Direct, accessed January 3, 2022 at www.sciencedirect.com/topics/biochemistry-genetics-and-molecular-biology/iridoid-glycosides

117. Natalie G. Mueller, "A Curious Quest to Rediscover Lost Crops," accessed December 1, 2017 at ngmueller.net/2017/09/21/a-curious-quest-to-rediscover-lost-crops/

118. Kindscher, *Edible Wild Plants of the Prairie,* 82.

119. Carol Gracie, *Spring Wildflowers of the Northeast,* (Princeton University Press, 2012), 165.

120. Kindscher, *Edible Wild Plants of the Prairie,* 89–90.

121. Porcher and Rayner, *A Guide to the Wildflowers of South Carolina,* 138.

122. Winston, *Nvwoti; Cherokee Medicine,* 3.

123. Haines, *Ancestral Plants Volume 1,* 68.

124. Shreya Kamath, Matthew Skeels, Aswini Pai, "Significant differences in alkaloid content of Coptis chinensis (Huanglian), from its related American species," *Chinese Medicine* 4, no. 1 (2009), 1–4.

125. Kindscher, *Edible Wild Plants of the Prairie,* 100.

126. Ibid., 101.

127. "Beltane Blessings," Green Man Ramblings, accessed March 3, 2019 at greenmanramblings.blogspot.com/2017/05/beltane-blessings.html

128. Lisa Ganora, *Herbal Constituents: Foundations of Phytochemistry,* (Herbalchem Press, 2009), 120.

129. Rosemary Gladstar, *Herbal Healing for Men: Remedies and Recipes,* (Storey Publishing, 2017), 179.

130. William H. Banks Jr., *Plants of the Cherokee,* (Great Smoky Mountains Association, 2004), 60.

131. "Crataegus," *Flora of North America,* Volume 9, accessed December 1, 2019 at www.efloras.org/florataxon.aspx?flora_id=1&taxon_id=108272

132. William Cullina, *Native Trees Shrubs and Vines: A Guide to Using, Growing, and Propagating North American Woody Plants,* (Houghton Mifflin, 2002), 284.

133. "Crataegus uniflora," Flora of North America, accessed December 1, 2019 at efloras.org/florataxon.aspx?flora_id=1&taxon_id=242416360

134. Phyllis D. Light, "Epigenetics & Methylation: The Influence of Our Ancestors," Lecture Handout, Allies for Plants and People Conference, 2018.

135. Ganora, *Herbal Constituents,* 143.

136. Ibid.

137. A. Pengelly and K. Bennett, "Appalachian plant monographs: Dioscorea villosa L., Wild Yam," accessed December 5, 2019 at www.frostburg.edu/aces/appalachian-plants/

138. Winston, *Nvwoti; Cherokee Medicine,* 3.

139. Porcher and Rayner, *A Guide to the Wildflowers of South Carolina,* 214.

140 Hill, *Weaving New Worlds,* 8.

141. Doolittle, *Cultivated Landscapes of Native North America.*

142. Gargiullo, *A Guide to the Native Plants of the New York City Region,* 8.

143. Nanci J. Ross, et al., "The ecological side of an ethnobotanical coin: Legacies in historically managed trees," *American Journal of Botany* 101.10 (2014), 1618–1630.

144. Patton, *Mountain Medicine,* 152.

145. B. J. Peterson, *Ecology and horticultural potential of Dirca palustris,* (Iowa State University, Graduate Thesis, 2009) 12.

146. Guy Sternberg and Jim W. Wilson, *Native Trees for North American Landscapes.* (Timber Press, 2004), 244.

147. R. R. Brooks, "Biological methods of prospecting for gold," *Journal of Geochemical Exploration* 17, no. 2 (1982), 109–122.

148. Kindscher, Yellow Bird, Yellow Bird, and Sutton, *Sahnish (Arikara) Ethnobotany,* 64.

149. Andrea Derksen, Joachim Kühn, Wali Hafezi, Jandirk Sendker, Christina Ehrhardt, Stephan Ludwig, and Andreas Hensel, "Antiviral activity of hydroalcoholic extract from Eupatorium perfoliatum L. against the attachment of influenza A virus," *Journal of Ethnopharmacology* 188 (2016), 144–152.

150. Buhner, *Herbal Antivirals,* 251.

151. Steven M. Colegate, Roy Upton, Dale R. Gardner, Kip E. Panter, Joseph M. Betz, "Potentially toxic pyrrolizidine alkaloids in Eupatorium perfoliatum and three related species. Implications for herbal use as boneset," *Phytochemical Analysis* 29, no. 6 (2018): 613–626.

152. Alfs, *Edible & Medicinal Wild Plants of the Midwest,* 57.

153. Matthew Wood, *The Earthwise Herbal: A Complete Guide to New World Medicinal Plants,* (North Atlantic Books, 2009), 154.

154. See for example the account in Matthew Wood, *The Book of Herbal Wisdom,* (North Atlantic Books, 1997).

155. Deborah G. McCollough, Robert L. Heyd, Joseph G. O'Brien, "Biology and Management of Beech Bark Disease," Michigan Extension Bulletin E-2746, accessed January 13, 2002 at migarden.msu.edu/uploads/files/e2746.pdf

156. Moerman, *Native American Medicinal Plants,* 204.

157. Porcher and Rayner, *A Guide to the Wildflowers of South Carolina,* 195.

158. Michael Kudish, *Adirondack Upland Flora: An Ecological Perspective,* (Chauncy Press, 1992), 118.

159. Kelly Kindscher, *Edible Wild Plants of the Prairie*, 117–118.

160. Ian Shackleford, "Creeping Snowberry (Gaultheria hispidula)," accessed January 13, 2022 at www.fs.fed.us/wildflowers/plant-of-the-week/gaultheria_hispidula.shtml

161. Ping Mao, Zizhen Liu, Meng Xie, Rui Jiang, Weirui Liu, Xiaohong Wang, Shen Meng, Gaimei She, "Naturally occurring methyl salicylate glycosides," *Mini Reviews in Medicinal Chemistry* 14, no. 1 (2014): 56–63.

162. Porcher and Rayner, *A Guide to the Wildflowers of South Carolina*, 251.

163. "Wild Geranium," Illinois Wildflowers, accessed January 13, 2022 at www.illinois wildflowers.info/woodland/plants/wild_geranium.htm

164. Henriette Kreiss, "Geranium. Geranium maculatum," accessed January 13, 2022 at www.henriettes-herb.com/eclectic/ellingwood/geranium.html

165. Patton, *Mountain Medicine*, 118.

166. Moerman, *Native American Medicinal Plants*, 216–217.

167. Aleksandra Owczarek, Jan Gudej, Monika Anna Olszewska, "Antioxidant activity of Geum rivale L. and Geum urbanum L.," L. Acta Pol Pharm 72 (2015): 1239–1244.

168. James A. Duke, "Gleditsia triacanthos," *Handbook of Energy Crops*, 1983, unpublished, accessed December 19, 2019 at www.hort.purdue.edu/newcrop/duke_energy/Gleditsia_triacanthos.html

169. Robert J. Warren, "Ghosts of Cultivation Past — Native American Dispersal Legacy Persists in Tree Distribution." *PloS one* 11, no. 3 (2016), 1.

170. Duke, *Handbook of Energy Crops*.

171. John B. Sullivan, Barry H. Rumack, Harold Thomas, Robert G. Peterson, Peter Bryson, "Pennyroyal oil poisoning and hepato-toxicity," *JAMA* 242, no. 26 (1979): 2873–2874.

172. "American Pennyroyal," Illinois Wildflower, accessed at www.illinois wildflowers.info/woodland/plants/am_pennyroyal.htm

173. Kindscher, Yellow Bird, Yellow Bird, and Sutton. *Sahnish (Arikara) Ethnobotany*, 70.

174. "July 17, 1805." Journals of The Lewis and Clark Expedition, lewisandclarkjournals.unl.edu/item/lc.jrn.1805-07-17

175. Kindscher, Yellow Bird, Yellow Bird, and Sutton, *Sahnish (Arikara) Ethnobotany*, 70.

176. Saurabh Sudha Dhiman, Xin Zhao, Jinglin Li, Dongwook Kim, Vipin C. Kalia, In-Won Kim, Jae Young Kim, and Jung-Kul Lee, "Metal accumulation by sunflower (Helianthus annuus L.) and the efficacy of its biomass in enzymatic saccharification," *PloS one* 12, no. 4 (2017).

177. Omena Bernard Ojuederie and Olubukola Oluranti Babalola, "Microbial and Plant-Assisted Bioremediation of Heavy Metal Polluted Environments: A Review," *International Journal of Environmental Research and Public Health* 14, no. 12 (2017), 1504.

178. Ninad P. Gujarathi, Bryan J. Haney, Heidi J. Park, S. Ranil Wickramasinghe, James C. Linden, "Hairy roots of Helianthus annuus: a model system to study phytoremediation of tetracycline and oxytetracycline." *Biotechnology Progress* 21, no. 3 (2005), 775.

179. Subhash Babu, D. S. Rana, G. S. Yadav, Raghavendra Singh, S. K. Yadav, "A review on recycling of sunflower residue for sustaining soil health," *International Journal of Agronomy* 2014 (2014).

180. A. A. P. M. Azania, et al., "Allelopathic plants. 7. Sunflower (Helianthus annuus L.)," *Allelopathy Journal*, 11(1, 2003), 1–4.

181. Roger Edwin Wilson, *Allelopathy as expressed by Helianthus annuus, and its role in oil-field succession* (Doctoral dissertation, The University of Oklahoma, 1968).

182. Tehmina Anjum, Phil Stevenson, David Hall, Rukshana Bajwa, "Allelopathic potential of Helianthus annuus L.

(sunflower) as natural herbicide," in *Proceedings of the Fourth World Congress on Allelopathy, WaggaWagga* (August, 2005) pp. 21–26.

183. Joanna Slavin, "Fiber and Prebiotics: Mechanisms and Health Benefits," *Nutrients*, vol. 5, no. 4, (2013), 1417.

184. Kindscher, *Edible Wild Plants of the Prairie*, 131.

185. Ibid., 132.

186. Ganora, *Herbal Constituents: Foundations of Phytochemistry*, 160.

187. Patton, *Mountain Medicine*, 131.

188. Guido Masé, *The Wild Medicine Solution: Healing with Aromatic, Bitter, and Tonic Plants*, (Simon and Schuster, 2013).

189. Moerman, *Native American Food Plants*, 129.

190. Simon A. Sarkisian, et al., "Inhibition of bacterial growth and biofilm production by constituents from Hypericum spp.," *Phytotherapy Research* 26, no. 7 (2012), 1012.

191. Doolittle, *Cultivated Landscapes of Native North America*, 65.

192. Ibid., 76.

193. Light, "Epigenetics & Methylation," 8.

194. Hill, *Weaving New Worlds*, 9.

195. Patton, *Mountain Medicine*, 99.

196. Light, "Epigenetics & Methylation," 8.

197. Gargiullo, *A Guide to the Native Plants of the New York City Region*, 9.

198. Mary Siisip Geniusz, *Plants Have So Much To Give Us, All We Have To Do Is Ask: Anishinaabe Botanical Teachings* (Minneapolis: University of Minnesota Press, 2015), 114.

199. "Juniperis communis," Plants for a Future, accessed January 13, 2022 at pfaf.org/user/Plant.aspx?LatinName=Juniperus+communis

200. Moerman, *Native American Medicinal Plants*, 251.

201. A. C. Whitford, *Textile Fibers Used in Eastern Aboriginal North America*. (American Museum of Natural History, 1941), 13.

202. Moerman, *Native American Medicinal Plants*, 262–263.

203. Daniel E. Moerman, "The medicinal flora of native North America: an analysis," *Journal of Ethnopharmacology* 31, no. 1 (1991): 1–42.

204. Weakley, *Flora of the Southeastern United States*, 808.

205. Haines, *Ancestral Plants Volume 2*, 58.

206. Joshua S. Caplan, et al., "Nutrient foraging strategies are associated with productivity and population growth in forest shrubs," *Annals of Botany* 119, no. 6 (2017): 978.

207. Mei-Li Zhao, et al., "Comparative chloroplast genomics and phylogenetics of nine Lindera species (Lauraceae)," *Scientific Reports* 8, no. 1 (2018), 1.

208. Moerman, *Native American Medicinal Plants*, 275.

209. David Winston, "American Extra Pharmocopeia," in Rosemary Gladstar and Pamela Hirsch, eds. *Planting the Future: Saving Our Medicinal Herbs*, (Inner Traditions/Bear & Co, 2000), 44.

210. Patton, *Mountain Medicine*, 194.

211. Gargiullo, *A Guide to the Native Plants of the New York City Region*, 10.

212. Spira, *Wildflowers and Plant Communities of the Southern Appalachian Mountains and Piedmont*, 242.

213. Patton, *Mountain Medicine*, 198.

214. Micah Dettweiler, et al., "American Civil War plant medicines inhibit growth, biofilm formation, and quorum sensing by multidrug-resistant bacteria," *Scientific Reports* 9, no. 1 (2019), 3.

215. Winston, *Nvwoti; Cherokee Medicine and Ethnobotany*, 6.

216. Moerman, *Native American Food Plants*, 148.

217. Thayer, *Nature's Garden*, 394.

218. Patton, *Mountain Medicine*, 107.

219. Moerman, *Native American Food Plants*, 151–152.

220. D. Fuller, "Effects of Long-Term Fiddlehead Harvest on Ostrich Fern, Matteuccia

struthiopteris," National Association of County Agriculture Agents 13, no. 1 (June 2020).

221. Moerman, *Native American Medicinal Plants,* 303.

222. Ibid., 312.

223. Paola Mattarelli, et al., "Chemical composition and antimicrobial activity of essential oils from aerial parts of Monarda didyma and Monarda fistulosa cultivated in Italy," *Journal of Essential Oil Bearing Plants* 20, no. 1 (2017), 76.

224. Wood, *The Book of Herbal Wisdom,* 361–378.

225. Kindscher, *Edible Wild Plants of the Prairie,* 151.

226. Kindscher, Yellow Bird, Yellow Bird and Sutton. *Sahnish (Arikara) Ethnobotany,* 81–82.

227. A. Scott Hauser, "Morella pensylvanica," in: Fire Effects Information System, [Online]. U.S. Department of Agriculture, Forest Service, Rocky Mountain Research Station, Fire Sciences Laboratory, accessed January 13, 2022 at www.fs.fed.us/database/feis/plants/shrub/morpen/all.html

228. Ibid.

229. Allen R. Place and Edmund W. Stiles, "Living Off the Wax of the Land." *The Auk* 109 (2, 1992), 340.

230. Hill, *Weaving New Worlds,* 9

231. Moerman, *Native American Medicinal Plants,* 316.

232. John Smith, *The General Historie of Virginia, New England and The Summer Isles* (Vol. I), Project Gutenberg eBook.

233. Green Deane, "American Lotus, Worth Getting Wet For," Eat The Weeds, accessed December 9, 2019 at www.eattheweeds.com/american-lotus-worth-getting-wet-for-2/

234. Porcher and Rayner, *A Guide to the Wildflowers of South Carolina,* 337.

235. John S. Newberry, *Food and Fiber Plants of the North American Indians* (1887), (Kessenger Publishing, 2010), 9.

236. Moerman, *Native American Food Plants,* 162.

237. Richard J. Medve and Mary Lee Medve, *Edible Wild Plants of Pennsylvania and Neighboring States,* (Penn State Press, 2010), 19.

238. Gargiullo, *A Guide to the Native Plants of the New York City Region,* 122.

239. Marc D. Abrams, "Tales from the Blackgum, a Consummate Subordinate Tree," *BioScience,* April 2007, 345.

240. Ibid.

241. Diane Stonemetz, "A review of the clinical efficacy of evening primrose," *Holistic Nursing Practice* 22, no. 3 (2008), 171–174.

242. Mary Anne Borge, "Evening Primrose," The Natural Web, accessed January 9, 2019 at the-natural-web.org/2016/02/26/evening-primrose/

243. Kindscher, *Edible Wild Plants of the Prairie,* 155.

244. Lucas C. Majure, Walter S. Judd, Pamela S. Soltis, Douglas E. Soltis, "Taxonomic revision of the Opuntia humifusa complex (Opuntieae: Cactaceae) of the eastern United States," *Phytotaxa* 290, no. 1 (2017), 5.

245. Ibid., 1-65.

246. Winston, *Nvwoti; Cherokee Medicine and Ethnobotany,* 6.

247. Ibid.

248. Kindscher, *Edible Wild Plants of the Prairie,* 159–160.

249. H. C. Harrison, et al., "Ginseng," *Alternative Field Crops Manual,* University of Wisconsin, accessed January 13, 2022 at www.hort.purdue.edu/newcrop/afcm/ginseng.html

250. Robby Gardner, "Scientists Say Ginseng Compound May Have Exercise and Aging Benefits," *Nutritional Outlook.* (July 30, 2018.)

251. Lauren Garcia Chance, "Conserving Medicinally and Culturally Significant Southeastern Plants," Southern Highlands Reserve News, accessed January 13,

2022 at southernhighlandsreserve.org/conserving-medicinally-and-culturally-significant-southeastern-plants/

252. Kristen Johnson Gremillion, "The development of a mutualistic relationship between humans and maypops (Passiflora incarnata L.) in the southeastern United States." *Journal of Ethnobiology* 9, no. 2 (1989), 135–136.

253. Ibid, 139.

254. Spira, *Wildflowers and Plant Communities of the Southern Appalachian Mountains and Piedmont,* 419.

255. Patton, *Mountain Medicine,* 147.

256. Notes from lecture by Phyllis D. Light, Allies for Plants and People Conference, 2018.

257. Haines, *Ancestral Plants Volume 2,* 209.

258. Winston, *Nvwoti; Cherokee Medicine and Ethnobotany,* 3.

259. Janne L. Jokinen and Arno Sipponen, "Refined spruce resin to treat chronic wounds: rebirth of an old folkloristic therapy," *Advances in Wound Care* 5, no. 5 (2016), 198–207.

260. Moerman, *Native American Medicinal Plants,* 356.

261. Haines, *Ancestral Plants Volume 1,* 50.

262. Patton, *Mountain Medicine,* 154.

263. Haines, *Ancestral Plants Volume 1,* 50.

264. Ibid.

265. Geniusz, *Plants Have So Much To Give Us,* 93.

266. Donald Culross Peattie, *A Natural History of North American Trees,* (Houghton Mifflin, 2007), 73.

267. Andrew M. Barton and Daniel J. Grenier, "Dynamics of Jack Pine at the Southern Range Boundary in Downeast Maine," *Canadian Journal Of Forest Research* 38, no. 4 (2008), 733–743.

268. Jennifer H. Carey, "Pinus banksiana," Fire Effects Information System, U.S. Department of Agriculture, Forest Service, Rocky Mountain Research Station, Fire Sciences Laboratory, accessed January 4, 2022 at www.fs.fed.us/database/feis/plants/tree/pinban/all.htm

269. Corey L. Gucker, "Pinus rigida," Fire Effects Information System, U.S. Department of Agriculture, Forest Service, Rocky Mountain Research Station, Fire Sciences Laboratory, accessed January 4, 2022 at www.fs.fed.us/database/feis/plants/tree/pinrig/all.html

270. Peattie, *A Natural History of North American Trees,* 66.

271. Jennifer H. Carey, "Pinus strobus," Fire Effects Information System, U.S. Department of Agriculture, Forest Service, Rocky Mountain Research Station, Fire Sciences Laboratory, accessed November 23, 2021 at www.fs.fed.us/database/feis/plants/tree/pinstr/all.html

272. Turner and von Aderkas, "Sustained by First Nations," 305.

273. Hank Becker, "American Mayapple Yields Anticancer Extract," USDA-ARS Research News, July 17, 2000, accessed January 13, 2022 at www.ars.usda.gov/news-events/news/research-news/2000/american-mayapple-yields-anticancer-extract/

274. John Mann, "Natural products in cancer chemotherapy: past, present and future," *Nature Reviews Cancer* 2, no. 2 (2002), 143.

275. Winston, *Nvwoti; Cherokee Medicine and Ethnobotany,* 3.

276. Jim McDonald, "Herbs for back and joint pain ...," Herbcraft, accessed January 3, 2022 at www.herbcraft.org/backpain.html

277. U.S. Department of Agriculture, "Dufourea novaeangliae," accessed November 3, 2019 at i5k.nal.usda.gov/dufourea-novaeangliae

278. Haines, *Ancestral Plants Volume 2,* 62.

279. Patton, *Mountain Medicine,* 134.

280. Kyung-Baeg Roh, Deokhoon Park, Eunsun Jung, "Inhibitory Effects of Prunella vulgaris L. Extract on 11-HSD1 in Human Skin Cells," *Evidence-Based Complementary and Alternative Medicine,* 2018.

281. Su-Juan Wang, et al., "Prunella vulgaris: a comprehensive review of chemical constituents, pharmacological effects and clinical applications," *Current Pharmaceutical Design* 25, no. 3 (2019), 359.

282. Yi-Xiang Hu, et al., "Antihepatofibrotic effects of aqueous extract of Prunella vulgaris on carbon tetrachloride-induced hepatic fibrosis in rats," *Planta medica* 82, no. 01/02 (2016), 97.

283. ChoonSeok Oh, et al., "Inhibition of HIV-1 infection by aqueous extracts of Prunella vulgaris L.," *Virology Journal* 8, no. 1 (2011), 1.

284. Porcher and Rayner, *A Guide to the Wildflowers of South Carolina*, 378.

285. Kindscher, *Edible Wild Plants of the Prairie*, 171.

286. Moerman, *Native American Medicinal Plants*, 386-387.

287. Doolittle, *Cultivated Landscapes of Native North America*, 72.

288. Ibid., 65.

289. V. Havard, "Food plants of the North American Indians," *Bulletin of the Torrey Botanical Club*, 22(3 1895), 103.

290. Kindscher, *Edible Wild Plants of the Prairie*, 175.

291. Patton, *Mountain Medicine*, 208.

292. Moerman, *Native American Food Plants*, 200.

293. Wood, *The Earthwise Herbal*, 286-291.

294. Patton, *Mountain Medicine*, 201-202.

295. Spira, *Wildflowers and Plant Communities*, 255.

296. Kim Eierman, "Powerful Prunus: A Visit With Dr. Doug Tallamy," accessed October 30, 2019 at www.ecobeneficial.com/2015/02/powerful-prunus-visit-dr-doug-tallamy/

297. Kindscher, *Edible Wild Plants of the Prairie*, 178.

298. Sean Sherman, *The Sioux Chef's Indigenous Kitchen*, (University of Minnesota Press, 2017), 173.

299. Jim McDonald, "Wild Cherry," accessed January 2, 2022 at www.herbrally.com/monographs/wild-cherry

300. Kiva Rose, "Cherry: At The Heart," accessed January 2, 2022 at enchantersgreen.com/cherry

301. Patton, *Mountain Medicine*, 166.

302. Winston, *Nvwoti; Cherokee Medicine and Ethnobotany*, 5.

303. Wood, *The Earthwise Herbal*, 170.

304. Veronika Valková, Hana Ďuranová, Lucia Galovičová, Nenad L. Vukovic, Milena Vukic, and Miroslava Kačániová, "In Vitro antimicrobial activity of lavender, mint, and rosemary essential oils and the effect of their vapours on growth of Penicillium spp. in a bread model system," *Molecules* 26, no. 13 (2021), 3859.

305. Wood, *The Earthwise Herbal*, 345-349.

306. William N. Setzer, Lam Duong, Trang Pham, Ambika Poudel, Cuong Nguyen, and Srinivasa Rao Mentreddy, "Essential Oils of Four Virginia Mountain Mint (Pycnanthemum virginianum) Varieties Grown in North Alabama," *Plants* 10, no. 7 (2021), 1397.

307. Micah Dettweiler, James T. Lyles, Kate Nelson, et al., "American Civil War plant medicines inhibit growth, biofilm formation, and quorum sensing by multidrug-resistant bacteria," *Scientific Reports*, May 22 2019, 5.

308. Porcher and Rayner, *A Guide to the Wildflowers of South Carolina*, 213.

309. Wood, *The Earthwise Herbal*, 293-296.

310. Hill, *Weaving New Worlds*, 11

311. Michael Kudish, *Adirondack Upland Flora: An Ecological Perspective*, (Chauncy Press, 1992), 120.

312. Kerry Hardy, *Notes on a Lost Flute: A Field Guide to the Wabanaki*, (Downeast Books, 2009), 67.

313. John D. Stein, *Field guide to native oak species of eastern North America*. Vol. 3, no. 1. US

Forest Service, Forest Health Technology Enterprise Team, 2003, 88.

314. "Rhodiola Root," accessed January 17, 2020 at www.mountainroseherbs.com/products/rhodiola-root-north-american/profile

315. Ibid.

316. Kindscher, *Edible Wild Plants of the Prairie*, 193.

317. Wood, *The Earthwise Herbal*, 298.

318. Patton, *Mountain Medicine*, 191.

319. Otis C. Maloy, "White Pine Blister Rust," accessed March 2, 2018 at www.apsnet.org/edcenter/intropp/lessons/fungi/Basidiomycetes/Pages/WhitePine.aspx

320. Glen Howard Mittelhauser, Marilee Lovit, and Donna Kausen, *Plants of Acadia National Park*, (University of Maine Press, 2010), 199.

321. "Ribes hirtellum," Native Plant Trust, accessed February 1, 2020 at https://gobotany.newenglandwild.org/species/ribes/hirtellum/

322. Lee Reich, *Uncommon Fruits for Every Garden*, (Timber Press, 2004), 158.

323. Michael Kudish, *Adirondack Upland Flora: An Ecological Perspective*, (Chauncy Press, 1992), 149.

324. Ann Fowler Rhoads and Timothy A. Block, *The Plants of Pennsylvania: An Illustrated Manual*, (University of Pennsylvania Press, 2007), 496.

325. "Ribes glandulosum," Native Plant Trust, accessed February 1, 2020 at gobotany.newenglandwild.org/species/ribes/glandulosum/

326. Jimmy C. Huntley, "Robinia pseudoacacia L. black locust," *Silvics of North America* 2 (1990), 755–761.

327. Katharine R. Stone, "Robinia pseudoacacia," Fire Effects Information System, [Online]. U.S. Department of Agriculture, Forest Service, Rocky Mountain Research Station, Fire Sciences Laboratory, accessed November 18, 2018 at www.fs.fed.us/database/feis/plants/tree/robpse/all.html

328. Karoly Redei, "Black locust (Robinia pseudoacacia L.) improvement and management in Hungary," *Forestry Studies in China* 1, no. 2 (1999), 42–46.

329. Stone, "Robinia pseudoacacia."

330. Susun W. Weed, *Wise Woman Herbal for the Childbearing Year*, (Ash Tree Publishing, 1986), 18–19.

331. Hoffman, *Holistic Herbal*, 227.

332. Kudish, *Adirondack Upland Flora*, 157.

333. Leigh Ann Henion, "Cherokee Indians Can Now Harvest Sochan Within a National Park," *Smithsonian*, October 2019.

334. Samuel Thayer, *Forager's Harvest: A Guide to Identifying, Harvesting, and Preparing Edible Wild Plants*, (Forager's Harvest Press, 2006), 115.

335. Patton, *Mountain Medicine*, 210.

336. Porcher and Rayner, *A Guide to the Wildflowers of South Carolina*, 174.

337. Kindscher, Yellow Bird, Yellow Bird, and Sutton, *Sahnish (Arikara) Ethnobotany*, 105.

338. Weed, *Wise Woman Herbal*, 117.

339. Buhner, *Herbal Antivirals*, (Storey Publishing, 2013), 144–149.

340. Patton, *Mountain Medicine*, 127–128.

341. Ibid., 177.

342. Wood, *The Earthwise Herbal*, 315–316.

343. Ganora, *Herbal Constituents: Foundations of Phytochemistry*, 109.

344. Elizabeth A. Hausner and Robert H. Poppenga, "Hazards Associated with the Use of Herbal and Other Natural Products," in *Small Animal Toxicology, Third Edition*, (Elsevier Inc., 2012), 335.

345. Traci Vogel, "Your Sassafras Has Been Neutered," Chowhound, accessed January 4, 2022 at www.chowhound.com/food-news/53525/your-sassafras-has-been-neutered/

346. Wood, *The Earthwise Herbal: New World*, 323-325.

347. Buhner, *Herbal Antivirals*, 125–127.

348. Ibid.

349. "Water Parsnip," Illinois Wildflowers, accessed January 11, 2022 at www.illinoiswildflowers.info/wetland/plants/wt_parsnip.html

350. Spira, *Wildflowers and Plant Communities*, 441.

351. Wood, *The Earthwise Herbal*, 468–469.

352. Huron H. Smith, "Ethnobotany of the Ojibwe Indians," *Bulletin Of The Public Museum Of The City Of Milwaukee* Vol. 4, No. 3, (May 2, 1932), 366.

353. Huron H. Smith, "Ethnobotany of the Meskwaki Indians," *Bulletin Of The Public Museum Of The City Of Milwaukee* VoL 4, No. 2 (April 7th, 1928), 218.

354. Geniusz, *Plants Have So Much To Give Us*, 38.

355. Ibid.

356. Haines, *Ancestral Plants Volume 1*, 61–62.

357. Janet Sullivan, "Tilia americana," Fire Effects Information System, [Online]. U.S. Department of Agriculture, Forest Service, Rocky Mountain Research Station, Fire Sciences Laboratory, accessed January 2 , 2022 at www.fs.fed.us/database/feis/plants/tree/tilame/all.html

358. Hoffman, *Holistic Herbal*, 211.

359. Michael Kudish, *Adirondack Upland Flora: An Ecological Perspective*, (Chauncy Press, 1992), 131.

360. P. A. Kulakow, M. K. Handley, J. Henson, and C. L. Dewald, "Comparison of Monoecious and Gynomonoecious Eastern Gamagrass Breeding Systems for Foliar Disease Susceptibility and Agronomic Traits," *Eastern Gamagrass Conference Proceedings*, (Poteau, Oklahoma January 23–25, 1989), 41.

361. Mittelhauser, Lovit, and Kausen, *Plants of Acadia National Park*, 398.

362. Geoffery A. Hammerson, *Connecticut Wildlife: Biodiversity, Natural History, and Conservation*, (University Press of New England, 2004), 139.

363. Haines, *Ancestral Plants Volume 1*, 65.

364. Patton, *Mountain Medicine*, 183–184.

365. Ibid.

366. Spira, *Wildflowers and Plant Communities*, 267.

367. Ibid., 318.

368. Eric Yarnell and Kathy Abascal, "Antiadhesion Herbs," *Alternative And Complementary Therapies* (2008), 139.

369. D. Tirmenstein, "Vaccinium vitis-idaea," Fire Effects Information System, U.S. Department of Agriculture, Forest Service, Rocky Mountain Research Station, Fire Sciences Laboratory, accessed January 5 2022 at www.fs.fed.us/database/feis/plants/shrub/vacvit/all.html

370. Patton, *Mountain Medicine*, 135.

371. W. Cassidy, A. Kalt, L. R. Howard, R. Krikorian, A. J. Stull, F. Tremblay, R. Zamora-Ros, "Recent research on the health benefits of blueberries and their anthocyanins," *Advances in Nutrition*, 11(2), 224.

372. Wood, *Earthwise Herbal*, 350.

373. D. Premilovac, K. M. Roberts-Thompson, H. L. H. Ng, E. A. Bradley, S. M. Richards, S. Rattigan, and M. A. Keske, "Blueberry tea enhances insulin sensitivity by augmenting insulin-mediated metabolic and microvascular responses in skeletal muscle of high fat fed rats," *International Journal of Diabetology & Vascular Disease Research* 1, no. 8 (2013), 1–10.

374. Spira, *Wildflowers and Plant Communities*, 317.

375. Jennifer Heinzel, "Discovering the Vervain Plant, Part 2," Mother Earth Living, accessed January 6, 2018 at www.motherearthliving.com/in-the-garden/discovering-the-vervain-plant-part-2

376. Wood, *The Earthwise Herbal*, 352.

377. Patton, *Mountain Medicine*, 103.

378. Wood, *The Earthwise Herbal*, 355–362.

379. "Viburnum lantanoides," Plants For a Future, accessed February 21, 2019 at pfaf.org/user/plant.aspx?latinname=Viburnum+lantanoides

380. Wood, *The Earthwise Herbal*, 361–362.

381. Patton, *Mountain Medicine*, 98.

382. "European Cranberrybush," EDDMaps, accessed January 13, 2021 at www.eddmaps.org/distribution/uscounty.cfm?sub=22400

383. Thayer, *Foragers Harvest*, 321.

384. "Viburnum opulus," Native Plant Trust, accessed March 3, 2018 at gobotany.newenglandwild.org/species/viburnum/opulus/

385. Wood, *The Earthwise Herbal*, 355–358.

386. Julia Blankenspoor, "Violet's Edible and Medicinal Uses," *Blog Castanea* (4/13/2016) accessed January 13, 2022 at chestnutherbs.com/violets-edible-and-medicinal-uses/

387. Ibid.

388. Spira, *Wildflowers and Plant Communities*, 467.

389. Doolittle, *Cultivated Landscapes of Native North America*, 62.

390. Ibid., 63.

391. Moerman, *Native American Food Plants*, 283.

392. Porcher and Rayner, *A Guide to the Wildflowers of South Carolina*, 327.

Chapter 11

393. Sai Gong, Chen Chen, Jingxian Zhu, Guangyao Qi, Shuxia Jiang, "Effects of wine-cap Stropharia cultivation on soil nutrients and bacterial communities in forestlands of northern China,", *PeerJ* 6 (2018), 5741.

394. Shivanand Hiremath, Kirsten Lehtoma, Jenise M. Bauman, "Native mycorrhizal fungi replace introduced fungal species on Virginia pine and American chestnut planted on reclaimed mine sites of Ohio," *Journal of American Society of Mining and Reclamation.* 3 (1). 1–15. 3, no. 1 (2014), 1–15.

395. Leslie Jones Sauer, *The Once and Future Forest*, (Island Press, 1998), 254–255.

Chapter 12

396. Christopher Haight, Sarah Lumban Tobing, Jessica A. Schuler, et al., "Japanese Knotweed Management in the Riparian Zone of the Bronx River," *Ecological Restoration* 35, no. 4 (2017), 298.

397. "Managing Japanese Knotweed and Giant Knotweed on Roadsides," Penn State, accessed January 11, 2022 at plantscience.psu.edu/research/projects/vegetative-management/publications/roadside-vegetative-mangement-factsheets/5 managing-knotweed-on-roadsides

Chapter 13

398. Marlin L. Bowles, Karel A. Jacobs, Jeffrey L. Mengler, "Long-term changes in an oak forest's woody understory and herb layer with repeated burning," *The Journal of the Torrey Botanical Society* 134, no. 2 (2007), 223.

399. Robin Wall Kimmerer and Frank Lake, "Maintaining the Mosaic: The role of Indigenous burning in land management," *Journal of Forestry*, (2001), 99. 36–38.

400. Ibid., 39.

401. Emily W. B. (Russell) Southgate, *People and the Land Through Time: Linking Ecology and History.* (Yale University Press), 1997, 81.

402. Kevin C. Ryan, Eric E. Knapp, J. Morgan Varner, "Prescribed fire in North American forests and woodlands: history, current practice, and challenges," *Frontiers in Ecology and the Environment* 11, no. s1 (2013), e17.

403. Jan W. Van Wagtendonk, "The history and evolution of wildland fire use," *Fire Ecology* 3, no. 2 (2007), 5.

404. Ryan, Knapp, Varner, "Prescribed fire in North American forests and woodlands," e17.

405. Ibid., e19.

406. Ibid., e15.

407. Matthew A. Albrecht and Brian C. McCarthy. "Effects of prescribed fire and thinning on tree recruitment patterns in central hardwood forests," *Forest Ecology and Management* 226, no. 1–3 (2006), 88

408. Marlin Bowles, Karel A. Jacobs, Jeffrey L. Mengler, "Long-term changes in an oak forest's woody understory and herb layer with repeated burning", *The Journal of the Torrey Botanical Society* 134, no. 2 (2007), 232.

409. Jay F. Kelly, Jessica Ray, Natale Minicuci, Rebekah Buczynski, "Effects of Prescribed Burning on Forest Understories in Northern New Jersey," Center for Environmental Studies, Raritan Valley Community College (2021), 12.

410. Ibid., 13.

411. "Fire Ecology and Indigenous Knowledge with Frank Lake," Good Fire Podcast (October 28, 2019).

Chapter 15

412. Leslie Sauer, email to author, November 16, 2021.

Conclusion

413. Adrian Barrance, Kathrin Schreckenberg, James Gordon, "Conservation Through Use: Lessons from the Mesoamerican Dry Forest," (Overseas Development Institute 2009), 3.

414. Kindscher, Yellow Bird, Yellow Bird, and Sutton, *Sahnish (Arikara) Ethnobotany*, 11–14.

415. Melvin Randolph Gilmore, *Uses of Plants by the Indians of the Missouri River Region,* (University of Nebraska Press, 1991), 53.

416. Tom Hatley, "Cherokee Women Farmers Hold Their Ground," in Gregory A. Waselkov, Peter H. Wood, and M. Thomas Hatley, eds. *Powhatan's Mantle: Indians in the Colonial Southeast,* (University of Nebraska Press, 2006), 308.

417. Ibid., 309.

418. Ibid., 307.

419. Ibid.

420. Charles W. Spellman, "The agriculture of the early north Florida Indians," *The Florida Anthropologist,* (1948), 44.

421. Spellman labels this *Smilax hastata.*

422. Hardy, *Notes on a Lost Flute*, 75–76.

423. Donald M. Waller and Nicholas J. Reo, "First stewards: ecological outcomes of forest and wildlife stewardship by indigenous peoples of Wisconsin, USA," *Ecology and Society* (2018), 45.

424. Ibid., 44.

425. Ibid., 45.

426. Ben Goldfarb, *Eager: The Surprising, Secret Life of Beavers and Why They Matter,* (Chelsea Green Publishing, 2018).

Index

Page numbers in *italics* indicate tables.

About the Author

JARED ROSENBAUM is a botanist, native plant grower, and ecological restoration practitioner. He is a founding partner at Wild Ridge Plants LLC, a business that grows local ecotype native plants using sustainable practices, performs botanical surveys, and provides ecological restoration planning services. Jared is the author of the children's book *The Puddle Garden*, about native plants and wildlife. He is a Certified Ecological Restoration Practitioner by the Society for Ecological Restoration. He loves music and plays dark post-punk, flamenco, and fingerstyle blues guitar. He lives in New Jersey.

ABOUT NEW SOCIETY PUBLISHERS

New Society Publishers is an activist, solutions-oriented publisher focused on publishing books to build a more just and sustainable future. Our books offer tips, tools, and insights from leading experts in a wide range of areas.

We're proud to hold to the highest environmental and social standards of any publisher in North America. When you buy New Society books, you are part of the solution!

At New Society Publishers, we care deeply about *what* we publish — but also about *how* we do business.

- Our books are printed on 100% **post-consumer recycled paper**, processed chlorine-free, with low-VOC vegetable-based inks (since 2002).

- Our corporate structure is an innovative employee shareholder agreement, so we're one-third employee-owned (since 2015)

- We've created a Statement of Ethics (2021). The intent of this Statement is to act as a framework to guide our actions and facilitate feedback for continuous improvement of our work

- We're carbon-neutral (since 2006)

- We're certified as a B Corporation (since 2016)

- We're Signatories to the UN's Sustainable Development Goals (SDG) Publishers Compact (2020–2030, the Decade of Action)

To download our full catalog, sign up for our quarterly newsletter, and to learn more about New Society Publishers, please visit newsociety.com

ENVIRONMENTAL BENEFITS STATEMENT

New Society Publishers saved the following resources by printing the pages of this book on chlorine free paper made with 100% post-consumer waste.

TREES	WATER	ENERGY	SOLID WASTE	GREENHOUSE GASES
56	4,400	23	190	24,000
FULLY GROWN	GALLONS	MILLION BTUs	POUNDS	POUNDS

Environmental impact estimates were made using the Environmental Paper Network Paper Calculator 4.0. For more information visit www.papercalculator.org

Certified B Corporation

new society PUBLISHERS
www.newsociety.com

FSC
MIX
Paper from responsible sources
FSC® C016245

SDG PUBLISHERS COMPACT